CARBON NANOTUBES FOR POLYMER REINFORCEMENT

CARBON NANOTUBES FOR POLYMER REINFORCEMENT

Peng-Cheng Ma
Leibniz Institute of Polymer Research,
Leibniz, Germany

Jang-Kyo Kim
Hong Kong University of Science and Technology,
Hong Kong SAR, China

CRC Press
Taylor & Francis Group
Boca Raton London New York

CRC Press is an imprint of the
Taylor & Francis Group, an **informa** business

CRC Press
Taylor & Francis Group
6000 Broken Sound Parkway NW, Suite 300
Boca Raton, FL 33487-2742

First issued in paperback 2017

© 2011 by Taylor and Francis Group, LLC
CRC Press is an imprint of Taylor & Francis Group, an Informa business

No claim to original U.S. Government works

ISBN-13: 978-1-4398-2621-8 (hbk)
ISBN-13: 978-1-138-07330-2 (pbk)

Visit the Taylor & Francis Web site at
http://www.taylorandfrancis.com

and the CRC Press Web site at
http://www.crcpress.com

Contents

Foreword

Since the early discovery of carbon nanotubes (CNTs) in the 1950s and recent advances made on their production methods, there have been tremendous research and development activities in the past decades to incorporate CNTs into polymer matrices in order to achieve superior physical and mechanical properties for a wide range of applications. However, two major obstacles stand in the way and they have to be resolved. One concerns the enhancement of interfacial adhesion between CNTs and the polymer matrix, and another is the improvement of dispersion of CNTs in the polymer. Thus, many efforts have been dedicated to these two issues and yielded promising results, which are reported in many publications scattered in different archive journals, workshop proceedings, and technical society meetings, making it difficult to access the full information and available knowledge in a single source.

Professor Jang-Kyo Kim and Dr. Peng-Cheng Ma have made substantial original contributions on the dispersion and functionalization of CNTs and their effects on the properties of CNT/polymer nanocomposites. Using mostly results of their own research studies, integrated with those of other investigators, they have written, in a single volume, the first monograph on the various conventional and innovative techniques available to disperse and functionalize carbon nanotubes for polymer reinforcement, elegantly explaining the basic sciences and technologies involved in those processes. Both structural and functional applications are also addressed in this book.

I have known Professor Kim for over 20 years. He is a full professor in the Hong Kong University of Science & Technology (HKUST), China, and is an outstanding researcher on fiber composites, especially composite interfaces on which he and I co-authored a monograph in 1998. Dr. Ma is an active researcher on composite materials and carbon nanotubes for functional applications. He is currently an Alexander von Humboldt Fellow affiliated with the Leibniz Institute of Polymer Research, Dresden, Germany.

CNT/polymer composites belong to that new class of *polymer nanocomposites*, which is gaining increasing importance and acceptance in many different applications. To this end, dispersion and functionalization of CNTs in polymers are critical processing parameters that must be precisely controlled. In this book the authors have given a comprehensive treatment and critical review of the subject, and pointed to areas of future development. Readers will find the basic knowledge and technical results contained herein informative and useful references for their work, whether this is for advanced research or for design and manufacturing of such composites. I can highly recommend it to composite engineers, scientists, researchers, and students. I have great confidence that this book will be a great success.

Yiu-Wing Mai
Sydney, Australia
August 2010

Preface

With unique structural and transport properties, carbon nanotubes (CNTs) have attracted much interest as the reinforcement for polymer matrix composites. CNT/polymer nanocomposites hold the promise of delivering exceptional mechanical properties and multifunctional characteristics. The potential of employing CNTs as reinforcements has, however, been severely limited because of difficulties associated with dispersion of entangled CNTs during processing and the poor interfacial interactions between CNTs and polymers. To ensure effective reinforcements for polymer composites, proper dispersion and good interfacial bonds between CNTs and polymers have to be guaranteed.

While there are a number of books that specialize in fabrication processes and properties of CNTs, very few books are available specifically dealing with the following topics for CNT application in polymer-based composites: (1) how CNTs are functionalized and their surface functionalities controlled to enhance the interfacial interactions with polymer matrix and (2) how CNTs are incorporated into polymers while the as-produced CNT agglomerates are dispersed into individual CNTs. In light of the authors' experience in CNT dispersion and functionalization in the past decades, this book is aimed at providing a comprehensive overview of conventional and novel techniques in dispersing and functionalizing CNTs for polymer reinforcements. It offers a systematic presentation of the principles, theories, and technical practices behind the dispersion of nanoparticles in general and the functionalization of the processes developed for CNTs, as well as the effects of CNT dispersion and functionalization on the resulting functional, mechanical, and structural properties of CNT/polymer nanocomposites.

The book is organized in five chapters. Chapter 1 gives an overview of CNTs and their properties. It also describes the methodology for characterization of CNTs, which are essential to understanding, optimizing, and exploiting CNT dispersion and functionalization processes for practical end applications. Chapter 2 introduces the principles and techniques for CNT dispersion. Principles behind the conventional and novel methods that are developed for CNT functionalization are presented in Chapter 3. The effects of functionalization on the properties of CNTs are discussed. Nanoparticle/CNT hybrid materials are particularly useful to integrate the properties of the two components for use in catalysis, energy storage, and other nanotechnologies. The latest development in this field is also discussed in Chapter 3. Chapter 4 is focused on the fabrication of CNT/polymer nanocomposites and the effects of CNT dispersion and functionalization on the properties of CNT/polymer nanocomposites. In-depth discussions are made on the role of the CNT–polymer interface on nanocomposite properties and behaviors, as well as on how to effectively control the interface from the viewpoint of nanocomposite design and specific applications. Chapter 5 is dedicated to practical applications of CNT/polymer nanocomposites.

The book is written primarily for an audience in the area of materials science and engineering in general, nanotechnology, carbon materials, polymer nanocomposites, and related fields, as well as for technical professionals, engineers, and scientists from the chemical, materials design and development, composite manufacturing, automotive, and aerospace industries. Nanotechnology, as an emerging new subject, has been established as a major or minor in undergraduate and postgraduate levels in many universities and research institutions around the globe. This book would be well suited as a text for an intermediate level course in such nanotechnology programs or graduate programs with a major in composites science and technology. It will be equally accessible to readers with either a science or engineering background. At the same time, the unique perspectives provided in the applications of CNTs and CNT/polymer nanocomposites will serve as a useful guide for design and fabrication of these composite materials for specific applications.

The authors are grateful for assistance and encouragement from many sources during the preparation of this book. They acknowledge the invaluable contributions of many colleagues and friends from different parts of the globe, with whom they had the privilege and delight to collaborate: including Professors B. Z. Tang, C. K. Y. Leung, G. Marom, S. H. Hong, K. W. Paik, A. Munir, E. Maeder, L. M. Zhou, R. K. Y. Li, and Y. W. Mai; as well as the research group members who carried out many of the research outputs quoted in this book.

Also greatly appreciated is the generous financial support provided by many organizations, most notably the Hong Kong University of Science & Technology (HKUST), which provided postgraduate studentships (PGS), Finetex Technology Global Ltd. through the Finetex-HKUST R&D Center at HKUST, and the Research Grant Council (RGC) of Hong Kong through various research grants, for performing the research recorded in this book. Thanks are also due to all those who have allowed the authors to reproduce photographs and diagrams from their published work and to their publishers for the permission to use them.

Peng-Cheng Ma and Jang-Kyo Kim
Clear Water Bay, Hong Kong
August 2010

Acronyms and Symbols

AC: alternating current
AFM: atomic force microscopy (or microscope)
Al₂O₃: alumina
ATRP: atom transfer radical polymerization
BAC: benzalkonium chloride
BMA: butyl methacrylate
BNNT: boron nitride nanotube
BZT: benzethonium chloride
C₆₀, C₇₀: fullerene
CB: carbon black
CMC: critical micelle concentration
CST: compression shear tests
CNF: carbon nanofiber
CNT: carbon nanotube
CO: carbon monoxide
CO₂: carbon dioxide
CPC: cetylpyridinium chloride
CTAB: cetyltrimethylammonium bromide
CVD: chemical vapor deposition
DC: direct current
DCC: di-cyclohexylcarbodiimide
DFT: density functional theory
DMA: dynamic mechanical analysis
DMAP: dimethylaminopyridine
DMF: N, N-dimethylformamide
DWCNT: double-walled carbon nanotube
EDX: energy dispersive x-ray spectrometer
EMI: electromagnetic interference
EP: epoxy
FEM: finite element method
FRP: fiber-reinforced polymer
FTO: fluorine doped tin oxide
GFRP: glass fiber-reinforced polymer
GNP: graphite nanoplatelet
HDPE: high-density polyethylene
HiPCO: high-pressure conversion of carbon monoxide
ILSS: interlaminar shear strength
INT: inorganic nanotube
IPA: isopropyl alcohol
IR: infrared spectroscopy

ITO: indium tin oxide
KB: potassium bromide
K_{IC}: quasi-static fracture toughness
LC: liquid crystal
LCP: liquid crystalline polymer
LIB: lithium ion batteries
MD: molecular dynamic simulation
MMA: methyl methacrylate
MS: mass spectrometry
MWCNT: multiwalled carbon nanotube
NEMS: nano-electro-mechanical system
NH_4HCO_3: ammonium bicarbonate
NL: nonlinear
NMP: N-methyl-2-pyrrolidone
NMR: nuclear magnetic resonance
NP: nanoparticle
OLED: organic light-emitting diode
OM: optical microscope
PA: polyamide
PANHS: pyrenebutyric acid N-hydroxysuccinimide ester
PB: polybutylene
PE: polyethylene
PEDOT: poly(3,4-ethylenedioxythiophene)
PEO: perfluorooctanoate
PET: polyethylene terephthalate
PFOS: perfluorooctanesulfonate
PI: polyimide
PL: photoluminescence
PLLA: poly(l-lactide)
PMMA: poly(methyl methacrylate)
PmPV: poly[(m-phenylenevinylene)-alt-(p-phenylenevinylene)]
PNA: peptide nucleic acid
POSS: polyhedral oligomeric silsesquioxane
PP: polypropylene
PPA: poly(phenylacetylenes)
PPY: polypyrrole
PS: polystyrene
PSS: poly(styrene sulfate)
PU: polyurethane
PVA: poly(vinyl alcohol)
PVC: poly(N-vinyl carbazole)
PVP: poly(vinylpyrrolidone)
RBM: radial breathing mode
RTM: resin transfer molding
SBS: short beam shear

SDBS: sodium dodecylbenzene sulfonate
SDS: sodium dodecyl sulfate
SE: shielding efficiency
SEM: scanning electron microscopy (or microscope)
SLES: sodium lauryl ether sulfate
SPM: scanning probe microscopy
SWCNT: single-walled carbon nanotube
TEM: transmission electron microscopy (or microscope)
TETA: triethylenetetramine
TGA: thermogravimetric analysis
UHMWPE: ultra-high molecular weight polyethylene
UV/O$_3$: ultraviolet/ozone
VR: vulcanized rubber
XPS: x-ray photoelectron spectroscopy
XRD: x-ray diffraction

a: unit vectors
d: diameter of spherical particle
E: Young's modulus
K: thermal conductivity
l: length of particle
m: slope
n **and** *m*: integers to describe the chirality of CNTs
s: conducting exponent
t: thickness of particle
V_f: filler volume fraction
V_c: percolation threshold

γ_s: total surface energy
γ_s^d: dispersive component of total surface energy
γ_s^p: polar component of total surface energy
γ_{LS}: interface energy or surface energy of interface
δ_t: total solubility parameter
δ_d: dispersive component of total solubility parameter
δ_p: high polar component of total solubility parameter
δ_h: hydrogen-bonding component of total solubility parameter
ε: strain
π **bond:** *Pi* bond
σ: stress or strength
σ: electrical conductivity
σ_0: electrical conductivity of the filler
σ **bond:** *Sigma* bond
ζ: Zeta potential
ω: radial breathing mode frequency of CNTs
ΔN_w **or** $\Delta G'$: *G'* band shift between zero strain and the applied strain ε of CNTs

1

Introduction

1.1 Introduction to Carbon Nanotubes (CNTs)

1.1.1 Family of Carbon Materials

Carbon, C, is a chemical element with an atomic number of 6. As a member of group 14 in the periodic table, it is a nonmetallic material. There are three naturally occurring isotopes of carbon, namely, ^{12}C, ^{13}C, and ^{14}C. The former two are stable, while the ^{14}C is a radioactive material with a half-life of about 5730 years. Carbon is one of the elements known since antiquity; the use of carbon black (CB) as a pigment in black ink can date back to the third century B.C.[1]. The word *carbon* originates from the Latin word *carbo*, and the word carbon can refer both to the element and to coal[2].

There are several allotropes of carbon, and the best known are graphite, diamond, fullerene (C_{60}, C_{70}, etc.), and CNTs. The macroscopic states of these carbon materials vary greatly with the allotropic form and structure, as shown in Figure 1.1. The structural differences result in large variations in physical and mechanical properties of the carbon allotropes.[1-5] For example, graphite, fullerene, and CNTs are opaque and black in appearance, while diamond is highly transparent. Diamond is among the hardest materials known, while graphite is soft enough to form a streak on paper—hence its name, from the Greek word "to write." Table 1.1 summarizes the electrical and thermal properties of carbon materials.[4,5]

All forms of carbon are highly stable because they are formed by tetravalent bonds; thus reactions of these materials require amenable environments, such as a high temperature. The most common oxidation state of carbon in inorganic compounds is +4, and +2 is found in carbon monoxide (CO) and some transitional metal carbonyl complexes, such as $Ni(CO)_4$, $Ti(CO)_7$, $Cr(CO)_6$, $Ru_3(CO)_{12}$, and so on. The largest sources of inorganic carbon are limestone (essentially $CaCO_3$), dolomite (or $CaMg(CO_3)_2$), and carbon dioxide (CO_2), but significant quantities occur in organic deposits, such as coal, peat, petroleum, and methane clathrates[2]. Carbon forms more compounds than any other element, and almost 10 million pure organic compounds have

1

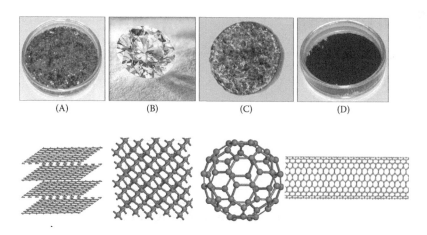

FIGURE 1.1
Macroscopic states and schematic structures of allotropes of carbon: (A) graphite, (B) diamond, (C) C_{60}, and (D) CNT.

TABLE 1.1   Physical Properties of Allotropes of Carbon.

Carbon Allotropes / Properties	Graphite	Diamond	Fullerene	CNT
Specific gravity (g/cm³)	1.9–2.3	3.5	1.7	0.8–1.8
Electrical conductivity (S/cm)	4000[p], 3.3[c]	10^{-2}–10^{-15}	10^{-5}	10^{2}–10^{6}
Electron mobility (cm²/(V·s))	~10^{4}	1800	0.5–6	10^{4}–10^{6}
Thermal conductivity (W/(m·K))	298[p], 2.2[c]	900–2320	0.4	2000–6000
Coefficient of thermal expansion (K⁻¹)	-1×10^{-6p}, 2.9×10^{-5c}	$(1\text{~}3)\times10^{-6}$	6.2×10^{-5}	Negligible
Thermal stability in air (°C)	450~650	<600	<600	>700

Note: p: in-plane; c: c-axis.
Source: Adapted from Kang, I., et al. 2006. Compos B 37: 382 and Ma, P.C., et al. 2010. Compos A 41: 1345.

been described to date, which are a tiny fraction of such compounds from the theoretical estimation.

Carbon plays an important role in the life cycle of creatures on Earth[3]. It is the 15th most abundant element in the Earth's crust, and the fourth most abundant element in the universe by mass after hydrogen, helium, and oxygen. It is present in all known life forms, and in the human body carbon is the second most abundant element by mass (about 18.5%) after oxygen (about 65.0%). The abundance, together with the unique diversity of organic compounds and their unusual ability to form polymers and biomacromolecules, make this element the chemical basis of all known life.

1.1.2 History of CNTs

The history of CNTs is not entirely clear even in the scientific communities[6–8]. The initial history of nanotubes started with the preparation of small-diameter carbon filaments through the decomposition of hydrocarbons at high temperatures in the presence of transition metal particles in the 1970s[9]. Although there are people who believe that these filaments are an initial discovery of CNTs, no detailed information of such thin filaments was reported in these early years, possibly due to the limitations on characterization and measurement. However, this is still an important development in the history of CNTs, as it triggered the interest of scientists to synthesize filament carbon materials that cover carbon fibers, CNTs, and carbon nanofibers.

A direct stimulus to the systematic study of carbon filaments with very small diameters arose from the discovery of fullerenes[10]. The experiments on the synthesis of new carbon clusters in the mid–1980s by Kroto and Smalley[10] are the prelude to the story of CNTs. From the evaporation of graphite, they discovered carbon clusters with different sizes and consisting of 60 and 70 carbon atoms. These unexpected findings brought fullerene into the spotlight, as schematically described in Figure 1.1C. Five years later, the success in the synthesis of fullerene in bulk quantity offered scientists an opportunity to study the properties of the new form of carbon[11].

In the early 1990s, Iijima investigated carbon soots and deposits produced by an electric arc discharge technique. By employing high-resolution transmission electron microscopy (TEM), he observed the existence of tube-like graphite structures with closed ends[12], which are now known as CNTs. From a historical point of view, if the anterior existence of CNTs is established, the importance of Iijima's work lies in the fact that it points out the exceptional properties and potential applications of these nanostructures, which heralds the entry of many scientists into the field of CNTs[6–8]. Whereas the initial observation by Iijima was multiwalled CNTs (MWCNTs), it was less than two years before single-walled CNTs (SWCNTs) were discovered independently by Iijima's group at the NEC Laboratory and by Bethune and coworkers at the IBM Almaden laboratory[13,14]. These findings are especially important because the SWCNTs are more fundamental, and have been the basis for a large body of theoretical studies and predictions.

Another major breakthrough in CNT history is the synthesis of aligned CNTs by Ebbesen and coworkers in 1993[15], thereby making it possible to carry out many sensitive experiments relevant to one-dimensional quantum physics, which could not be undertaken previously.

1.1.3 Synthesis of CNTs

Nowadays, CNTs are produced by three main techniques—electric arc discharge, laser ablation, and chemical vapor deposition (CVD)[16–19].

The arc discharge technique (Figure 1.2A) involves the generation of an electric arc between two graphite electrodes on which a voltage of 20–25 V

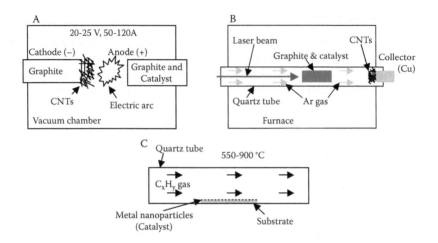

FIGURE 1.2
Schematics of the techniques to synthesize CNTs: (A) electric arc discharge, (B) laser ablation, and (C) chemical vapor deposition. Adapted from Thostenson, E. T., et al. 2001. *Compos. Sci. Technol.* 61: 1899; Hu, Y., et al. 2006. *Rep. Prog. Phys.* 69: 1847; Chen, Y., et al. 2006. *Sci. Technol. Adv. Mater.* 7: 839; and Wilson, M. 2002. *Nanotechnology: Basic science and emerging technologies*, 100. Boca Raton, FL: Chapman & Hall/CRC.

and a direct current of 50–120 A are applied. The anode is usually filled with a catalyst metal powder, such as Fe, Ni, and Co. CNTs can be collected from the cathode. This method can easily produce straight and near-perfect MWCNTs.

Laser ablation was first used for the synthesis of fullerene. Over the years, the technique has been improved to produce CNTs. In this technique (Figure 1.2B), a pulsed laser is used to vaporize a composite consisting of graphite and a catalyst (typically Co/Ni alloy) held in a furnace at a temperature near 1200°C. SWCNTs with high purity can be obtained by employing this method. Compared with MWCNTs prepared in an arc discharge chamber, the MWCNTs grown in a laser-ablation process are relatively short.

CVD is a process in which gaseous molecules, called *precursors*, are transformed into solid materials on a substrate at a range of temperatures between 550 and 900°C (Figure 1.2C). In this process, the catalytic particles deposited on the substrate can decompose hydrocarbon gas to form carbon and hydrogen, whereas the carbon dissolves into particles and precipitates out from its circumference to form CNTs. The catalyst acts as a template on which CNTs can grow, and by controlling the catalyst size and the reaction time, the diameter and length of CNTs can be controlled. Different hydrocarbon compounds, such as methane, ethylene, acetylene, benzene, toluene, and xylene, have been employed as precursors to produce CNTs. When CO is used as a carbon source at high temperature and pressure, gas-phase growth SWCNTs can be obtained. This improved CVD process is specifically called the *high-pressure conversion of carbon monoxide* (HiPCO) method. The CVD-grown CNTs usually show poor crystallinity and a postsynthesis at a high temperature is often applied to improve the product quality. The unique properties of CVD are: (1) the ability to synthesize well-aligned CNT arrays on a large scale, and (2) the possibility for selective CNT growth with controlled structures, such as nitrogen-doped CNTs with a bamboo-like structure. As a result, CVD is especially appealing in making electronic devices that require controlled growth of CNTs on prepatterned substrates.

It should be noted that the methods described herein are still under development, and there are numerous variations of these techniques operating under different conditions, with different experimental setups and processing parameters. Every technique provides unique advantages and disadvantages over the quality and kinds of synthesized CNTs. An overview of these techniques is given in Table 1.2.

Besides the aforementioned methods for the synthesis of CNTs, traditional techniques are also available to produce CNTs. For example, ball milling and subsequent annealing[18]. This method consists of placing graphite powder into a stainless steel container with steel balls. Milling is carried out at room temperature for up to 150 hours. The milled powder is annealed under an inert gas flow at 1400°C for 6 hours. The mechanisms for producing CNTs through this process are not known, but it is thought that the ball milling process forms nanotube nuclei, and the annealing process activates nanotube growth. This method produces more MWCNTs and few SWCNTs.

CNTs can also be produced by diffusion flame synthesis, electrolysis, use of solar energy, heat treatment of a polymer, and low-temperature solid pyrolysis[17,19]. It should be noted that no matter which synthesis technique is employed, the as-produced CNTs usually contain some impurities, such as catalyst particles and amorphous carbon nanoparticles. Several methods are available for purification, including oxidation in air, acid, and UV/ozone treatment, microfiltration and size exclusion chromatography, and so on. Examinations of the CNTs purified using chemical methods often show some structural changes, which in turn affect the CNT properties.

1.2 Properties of CNTs

1.2.1 Structure of CNTs

Unlike other carbon materials, such as graphite, diamond, and fullerene, CNTs are a one-dimensional carbon material that can have an aspect ratio greater than 1000. They can be envisioned as cylinders composed of rolled-up graphene planes with diameters in nanometer scale[20]. The cylindrical nanotube usually has at least one end capped with a hemisphere of fullerene structure.

TABLE 1.2　Comparison of Most Commonly Used Techniques for CNT Synthesis.

Method / Item	Arc Discharge		Laser Ablation	CVD	HiPCO
Principle	Generation of arc discharge between two graphite electrodes under an inert atmosphere (Ar, He).		Vaporization of graphite by laser under flowing inert atmosphere and high temperature.	Decomposition of hydrocarbon gases in the presence of metal catalyst particles.	Gas-phase growth of CNTs with CO at high pressure and temperature.
Dimension of CNTs — SWCNTs	Dia.=1–2 nm, short		Dia.=1–2 nm, long	Dia.=0.5–4 nm, long	Dia.< 1 nm, various length
Dimension of CNTs — MWCNTs	Dia.=1–10 nm, short		Dia.=1–20 nm, short	Dia.=10–200 nm, long	Not applicable
CNT Yield	Up to 90%		Up to 55%	Up to 100%	Up to 70%
Advantage	Easy, defect free, no catalyst		High purity, defect free	Low cost, simple, large scale production	Large scale, high purity
Disadvantage	Short, tangled CNTs, random structures		Expensive, low yield, for SWCNT only	Defects, catalyst and amorphous C particles, limited control over the CNT structures	Expensive, for SWCNT only

Source: Adapted from Hu, Y., et al. 2006. *Rep. Prog. Phys.* 69: 1847; Chen, Y., et al. 2006. *Sci. Technol. Adv. Mater.* 7: 839; and Wilson, M. 2002. *Nanotechnology: Basic science and emerging technologies,* 100. Boca Raton, FL: Chapman & Hall/CRC.

Depending on the process used for CNT synthesis, CNTs can be classified into single-walled and multiwalled carbon nanotubes (SWCNTs and MWCNTs). SWCNTs consist of a single graphene layer (Figure 1.3A), whereas MWCNTs consist of two or more concentric cylindrical shells of graphene sheets (Figure 1.3B) coaxially arranged around a central hollow core with van der Waals forces between adjacent layers.

According to the rolling angle of the graphene sheet, CNTs have three chiralities: armchair, zigzag, and chiral. The tube chirality is defined by the chiral vector, $C_h = na_1 + ma_2$ (Figure 1.4), where the integers (n, m) are the number of steps along the unit vectors (a_1 and a_2) of the hexagonal lattice[16,21].

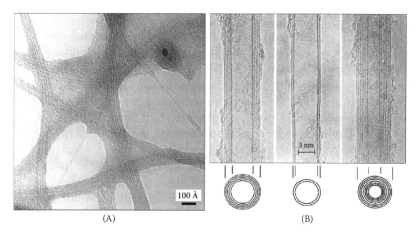

FIGURE 1.3
TEM images of different CNTs: (A) SWCNTs and (B) MWCNTs with varying number of graphene layers 5, 2, and 7. Adapted from Krätschmer, W., et al. 1990. *Nature* 347: 354 and Bethune, D. S., et al. 1993. *Nature* 1993 363: 605.

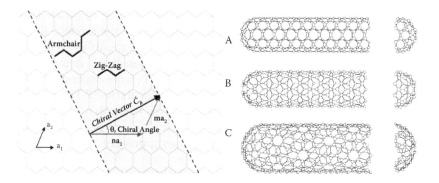

FIGURE 1.4
Schematic of how a hexagonal sheet of graphene is rolled to form a CNT with different chiralities: (A) armchair, (B) zigzag, and (C) chiral. Adapted from Thostenson, E. T., et al. 2001. *Compos. Sci. Technol.* 61: 1899 and Dresselhaus, M. S., et al. 1995. *Carbon* 33: 883.

Using this (*n, m*) naming scheme, three types of orientation of carbon atoms around the nanotube circumference are specified. If $n = m$, the nanotubes are called *armchair*. If $m = 0$, the nanotubes are called *zigzag*. Otherwise, they are called *chiral*. The chirality of nanotubes has significant impact on their transport properties, particularly the electrical and thermal conductivities that will be discussed in Sections 1.2.3 and 1.2.4.

1.2.2 Mechanical Properties of CNTs

The mechanical properties of CNTs are closely related to the nature of the chemical bonds between the carbon atoms. Since CNTs can be envisaged as a rolled-up graphene sheet with a single or multiple layers, their bonding mechanism is similar to that of graphite (or graphene for SWCNTs). A carbon atom consists of six electrons outside the nucleus, and its electronic structure can be expressed as $1s^2 2s^2 2p^2$. When carbon atoms are combined to form graphite, one *s* orbital and two *p* orbitals are combined to form three hybrid sp^2 orbitals at 120° to each other within a plane. The in-plane bond is known as a *sigma* bond (σ bond), a strong covalent bond with a bond energy of 346 KJ/mol, which is responsible for the high stiffness and high strength of a CNT. The remaining *p* orbital is perpendicular to the plane of the σ bond, contributing mainly to the interlayer interaction. It is called a *pi* bond (π bond) and is much weaker than the σ bond.

Practical applications of CNTs require a full understanding of their mechanical properties, including elastic and inelastic behavior, strength and fracture toughness, and myriad efforts have been directed toward theoretical and experimental investigations of these properties. In the early theoretical studies of the elastic moduli of CNTs, the nanotubes were modeled as a solid or hollow homogeneous cylinder[22,23]. In most of these models, the second derivative of the strain energy was calculated with respect to the nanotube elongation to derive the Young's modulus. The Young's moduli varied from around 0.8 TPa to as high as 5.5 TPa. This large variation originated from the ambiguity of nanotube type and diameter used in the models. Further study[24] indicated that SWCNTs could have a Young's modulus comparable to that found in diamond, which is about 1.0 TPa. Gao et al.[25] demonstrated theoretically that the mechanical properties of CNTs were dependent on the diameter, and their Young's modulus is in the range of 0.6–0.7 TPa when the diameter was less than 1 nm. This value was much lower than the 1–1.2 TPa for MWCNTs reported by Hernández et al.[26], who also proposed that as the diameter increased, their mechanical properties approached those of planar graphite.

Table 1.3 summarizes the experimental measurements of Young's modulus of various CNTs. The first attempt to experimentally measure the Young's modulus of MWCNTs was performed by Treacy and coworkers by monitoring the amplitudes of thermal vibrations of an individual MWCNT inside a TEM chamber[27]. An average Young's modulus of 1–1.8 TPa was reported,

TABLE 1.3 Mechanical Properties of CNTs Measured by Different Techniques.

	CNT Fabrication	Testing Method	Young's Modulus (TPa)	Strength (GPa)	Reference
SWCNTs	Spark plasma sintering	Three-point bending test	8.3	54	31
	Laser ablation	Thermal vibrations of CNT bundles in TEM	1.25	—	32
	Arc discharge	Direct measurement of CNT bundles using AFM tip	1.0	—	33
	Arc discharge	Direct measurement using AFM tip	0.32–1.47	13–52	34
	Arc discharge	Thermal vibration of CNTs in TEM	1–1.8	—	27
	Arc discharge	Direct measurement using AFM	1–1.28	—	30
MWCNTs	Arc discharge	Stress-strain behavior of individual CNTs in TEM	0.27–0.95	11–63	28
	Laser ablation	Direct measurement using a piezoelectric device in TEM	0.81	—	32
	CVD method	Stress-strain behavior of CNT bundle in TEM	0.45	4	34

which is much higher than those of commercially available carbon fibers, 200–350 GPa. Other research groups performed direct measurements in bending inside an atomic force microscopy (AFM). The reported values of Young's modulus fluctuated between 0.27 and 1.8 TPa. Decisive measurements were carried out by mounting an MWCNT on two AFM tips (Figure 1.5A)[28]. The stress-strain behavior of an individual MWCNT was obtained by stretching the two AFM tips (Figure 1.5C), which reported modulus values of 0.27–0.95 TPa for a range of CNTs. More interestingly, these MWCNTs exhibited failure strains of up to 12% and tensile strengths in the range of 11–63 GPa, allowing the estimation of nanotube toughness at about 1240 J/g. In addition, failure in MWCNTs was observed at the outer tube with the inner walls telescoping out in a "sword and sheath" mechanism. The large variations in the mechanical properties of CNTs arise from different qualities of CNTs used in these studies, which depended on several factors, including the crystallinity of the material, the number of defects present in the structure (e.g., pentagon-heptagon pairs, vacancies, interstitials, etc.), and the chiralities of CNTs.

Measurement of mechanical properties of SWCNTs is generally more difficult than measuring those of MWCNTs, partly because the as-produced SWCNTs are held together in bundles by van der Waals forces and to separate them into individual SWCNTs is untrivial. The first measurement was reported by Salvetat et al. using the AFM method on bundled SWCNTs[33], reporting a tensile modulus of about 1.0 TPa for small-diameter SWCNT bundles. The properties of large-diameter SWCNT bundles were

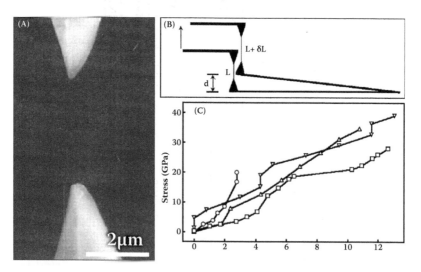

FIGURE 1.5
Measurement of mechanical properties of CNTs in a tensile test: (A) MWCNT mounted between two AFM tips, (B) schematic showing the principle of tensile test, and (C) stress-strain curves of MWCNTs. Adapted from Yu, M. F., et al. 2000. *Science* 287: 637.

dominated by shear slippage of individual nanotubes within the bundle. This was confirmed by measuring the tensile properties of SWCNT bundles by monitoring the stress-strain curves[34]. The Young's modulus and strength were reported to be in the range of 0.32–1.47 TPa and 13–52 GPa, respectively. Failure occurred at a maximum strain of 5.3%, thus giving a toughness of about 770 J/g. It is worth noting that failure occurred on the perimeter of the bundle while the rest of the tubes slipped apart.

1.2.3 Electrical Properties of CNTs

Section 1.2.1 indicates that the p orbital of electrons in carbon atoms forms the π bond in the CNT structure. Although the π bond is much weaker than the σ bond, it plays an important role in determining the electrical properties of CNTs. The π bond consists of a delocalized π electron that is free to move along the entire structure, rather than to remain within the donor atom, thus giving rise to unique electrical properties of CNTs. The electrical conductivity of a CNT is determined by the chirality (i.e., the degree of twist). The reason is that the electrons of carbon atoms are confined in the singular plane of the graphene sheet in the radial direction. The conduction occurs in the armchair tubes through gapless modes because the valence and conduction bands always cross each other at the Fermi energy, showing a metallic behavior of CNTs. However, for the helical tubes with different chiralities, there are a large number of atoms in their unit cell, and the one-dimensional band structure shows an opening of the gap at the Fermi energy, thus leading to semiconducting properties[35]. For a given (n, m) nanotube, if $(n - m)$ is a multiple of 3, then the nanotube is metallic, otherwise the nanotube is a semiconductor[36]. In theory, metallic nanotubes can carry an electrical current density of 4×10^9 A/cm^2, which is more than 1000 times greater than metals such as copper[37]. Therefore, the electrical conductivity of CNTs depends strongly on the synthesis method, which in turn affects the chirality and defect numbers of CNTs.

There has been considerable interest in the measurement of electrical properties of CNTs due to their potential applications in nanoscale electronic devices. Direct two- or four-probe measurements on individual CNTs can avoid the contact problems, thus these techniques have been commonly employed to determine the conductivity mechanism of CNTs (Figure 1.6). Tans and coworkers[38] have made the first attempt to measure the transport properties of individual SWCNTs using the two-probe method (Figure 1.6A). They demonstrated that SWCNTs indeed acted as genuine quantum wires, in which electrical conduction occurred via well-separated, discrete electron states that were quantum-mechanically coherent over long distances. The conductivity of SWCNT bundles has been measured by placing electrodes at different parts of the CNTs, and an electrical conductivity in the order of 10^4 S/cm at room temperature[40,41] has been reported. These reports suggested that even the bundle form of SWCNTs exhibits much higher conductivity than the most conductive carbon fibers (with a conductivity of about 100 S/cm).

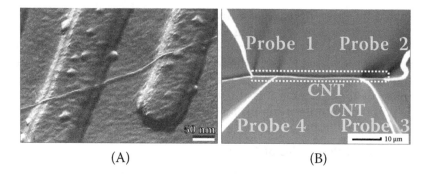

(A)   (B)

FIGURE 1.6

Electrical conductivity measurement of CNTs using different methods: (A) two-probe method for SWCNT and (B) four-probe method for MWCNT. Adapted from Tans, S.J., et al. 1997. *Nature* 386: 474 and Kim, T. H., et al. 2008. *Nanotechnology* 19: 485201.

Compared with SWCNTs, the electrical properties of MWCNTs are quite complex, as each MWCNT contains a multilayer of graphene, and each layer can have different chiralities. Furthermore, the interactions between the graphene layers within a MWCNT were found to disturb the current along the tube axis[42]. In the first electrical conductance measurement of an individual MWCNT, Langer et al.[43] found that the conductance of MWCNTs measured without a magnetic field exhibited a logarithmic decrease with decreasing temperature, and showed saturation when the temperature was below 0.3 K. In the presence of a magnetic field applied perpendicular to the tube axis, pronounced and positive magnetoresistance was observed. The temperature dependence of the conductance in a magnetic field was found to be consistent with two-dimensional weak localization. Subsequent studies demonstrated that each MWCNT exhibited a unique conductivity that can lead to both metallic and semiconducting behaviors, and the conductivity could increase a remarkable 10^4–10^6 S/cm at room temperature with activation energies less than 0.3 eV for semiconducting tubes[44]. These results suggest that both the geometric variations of CNTs (e.g., defects, chirality, number of graphene layers, and tube diameter) and the degree of crystallinity (i.e., hexagonal lattice perfection) affected by the CNT fabrication process played a key role in determining the electrical response of CNTs. The electrical conductivities measured of aligned MWCNTs[44-47] showed the materials behaving as conducting rods and having strong anisotropic properties for different alignment configurations.

1.2.4 Thermal Properties of CNTs

It is well known that the thermal conductivities of other carbon allotropes, such as diamond and graphite (in-plane direction), are extremely high (see Table 1.1), thus leading to the expectation that the longitudinal thermal conductivity of CNTs could possibly be comparable to the in-plane thermal conductivity of graphite. Berber et al.[48] predicted the thermal conductivity of individual SWCNTs to be as high as 6600 W/m·K at room temperature, exceeding the reported thermal conductivity of pure diamond by almost a factor of 2. Table 1.4 summarizes the thermal conductivities of various CNTs with different physical states, in either individual or CNT bundles. Apparently, these values measured of individual CNTs are often above 1000 W/m·K, which is significantly higher than those of highly conductive metals, such as silver, copper, and gold with their highest conductivity of about 430 W/m·K.

The thermal conductivity of a material is determined by the transportation of phonons. In graphite, phonons dominate the thermal conductivity at temperatures above 20 K, and it increases markedly as the temperature decreases[49]. In CNTs, however, the phonon contribution dominates at all temperatures[29]. Indeed, the thermal conductivity of both SWCNTs and MWCNTs varied linearly with temperature from 50 to 300 K[50,52], as shown in Figure 1.7. Because of the large diameter of MWCNTs, they acted essentially as two-dimensional phonon materials. Meanwhile, at low temperatures below 100 K, the thermal conductivity tended to increase as a function of temperature

TABLE 1.4   Thermal Conductivity of CNTs at Room Temperature.

	Physical State of CNTs	Testing Method	Thermal Conductivity (W/(m·K))	Reference
SWCNT	Theoretical analysis for (10, 10) CNT	Molecular dynamics simulation	6600	48
	Arc discharge CNT mat	Comparative method	35	54
	Aligned CNT thin film	Laser flash	~210	52
	Random CNT thin film	Laser flash	~20	52
	Individual CNT (l = 2.6 μm, d = 1.7 nm)	Extraction from I-V data	3500	56
MWCNT	CVD fabricated CNT bundle	Self-heating 3ω method	25	50
	Individual CNT (d = 14 nm)	Microfabricated device	3000	51
	CNT bundle (d = 80 nm)	Microfabricated device	~1300	51
	CNT bundle (d = 200 nm)	Microfabricated device	~300	51
	Bulk CNTs	Laser flash	4.2	57
	Individual CNT (d = 10 nm)	Electrical thermometry	600	58

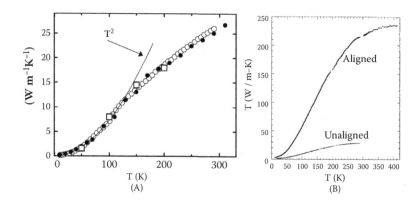

FIGURE 1.7
Thermal conductivity of CNTs as a function of temperature: (A) MWCNT bundles and (B) SWCNT bundles with different orientations. Adapted from Yi, W., et al. 1999. *Phys. Rev. B* 59: R9015 and Hone, J., et al. 2000. *Appl. Phys. Lett.* 77: 666.

squared (T^2). However, the conductivity measured at room temperature was relatively low, about 25 W/m·K, and comparable to the less-graphitic carbon fibers, possibly due to the small crystal size in the graphitic domains. In contrast, a completely different result was reported by Kim et al.[51] showing that the thermal conductivity of a single MWCNT with a diameter of 14 nm was about 3000 W/m·K. This value is much higher than that of graphite at room temperature, and two orders of magnitude higher than those obtained for MWCNT mats. The temperature dependence of thermal conductivity exhibited a peak at 320 K due to the onset of phonon scattering. They suggested that there were numerous highly resistive thermal junctions between the tubes in the mat samples.

Thess and coworkers carried out thermal conductivity measurements on bulk SWCNTs using a laser flash technique[53]. The results revealed temperature dependence of thermal conductivity, which is different from that of MWCNTs, due to the presence of smaller crystalline sizes within the graphitic domains of SWCNTs than those of MWCNTs. Hone et al.[54] further confirmed the linear dependence of thermal conductivity of SWCNTs on temperature at a low range, reflecting the one-dimensional band structure of individual SWCNTs in which linear acoustic bands contribute to thermal transport at low temperatures and optical sub-bands enter at high temperatures[55]. Pop et al.[56] investigated the thermal properties of a single SWCNT by extracting its high-bias ($I-V$) electrical characteristics at 300–800 K. The thermal conductivity was reported as a remarkable 3500 W/m·K at room temperature for a SWCNT with a length of 2.6 μm and a diameter of 1.7 nm. A subtle decrease in thermal conductivity steeper than 1/T was observed at the upper end of the temperature range, which is attributed to second-order three-phonon scattering between the two acoustic modes and one optical mode.

Besides the excellent thermal conductivities, CNTs also exhibit remarkable thermal stability. Figure 1.8 shows the thermogravimetric analysis (TGA) of different carbon materials under air flow[59]. It is clear that CNTs are far more resistant to oxidation than either graphite or C_{60} as indicated by the onset temperature for weight loss as well as the peak temperature corresponding to maximum oxidation rate. The peak temperature was 695°C for CNTs, which was much higher than that of C_{60} (420°C), graphite (645°C), and diamond (630°C)[60]. More significantly, TGA studies of CNTs in argon showed no weight change or detectable thermal transformation at temperatures up to 1000°C, indicating no bond breakage by pyrolysis or change in molecular packing. Note that the CNTs used in this study contained carbon nanoparticles that may have had negative effects on their thermal stability. It is estimated that the temperature stability of purified CNTs in vacuum can be as high as 2800°C[61]. The exceptionally high thermal stability and the excellent thermal conductivities of CNTs make them ideal materials for various thermal management applications.

1.2.5 Optical Properties of CNTs

The optical properties of CNTs refer to light absorption, photoluminescence (PL), and Raman scattering. These properties are important because they offer the possibility of nondestructive characterization of relatively large amounts of CNTs, thus allowing quick and reliable characterization of the quality of CNTs in terms of nontubular carbon content, chirality, and structural defects[62]. These features determine almost all of the important properties of CNTs, including mechanical, electrical, and thermal properties. Therefore, there has been strong interest in studying the optical properties of CNTs from both the academic and practical points of view.

The optical absorption in CNTs differs from absorption in conventional three-dimensional materials in view of the presence of sharp peaks in CNTs, instead of an absorption threshold followed by an absorption increase in the latter materials. Absorption in CNTs originates from electronic transitions from valence bands to conduction bands[63]. The transitions are relatively sharp and can be used to identify nanotube types. For example, for SWCNTs, the peaks located at 154, 157, 163, 178 cm^{-1} are assigned as (16, 4), (18, 0), (10, 10), (12, 6) CNTs, respectively. Figure 1.9A shows a typical optical absorption of purified CNTs. The peaks at around 0.6 eV and 1.2 eV correspond to metallic and semiconducting CNTs, respectively, whereas the large absorption band at 4.5 eV is ascribed to the π-plasmon of CNTs[64].

A strong light absorbance of 0.98–0.99 from the far-ultraviolet (200 nm) to far-infrared (200 μm) wavelengths has been reported for vertically aligned SWCNT forests or buckypaper fabricated by a CVD method[65]. This value is comparable to that of super black, a coating based on a chemically etched nickel-phosphorus alloy with an approximate absorbance of 1.0. Two factors governed the strong light absorption: (1) various band gaps of different CNTs

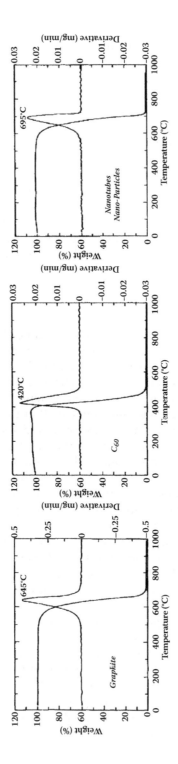

FIGURE 1.8

Thermogravimetric analysis of different carbon materials: (A) graphite, (B) C_{60} and (C) CNT. Adapted from Pang, L. S., et al. 1993. *J. Phys. Chem.* 97: 6941.

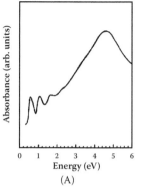 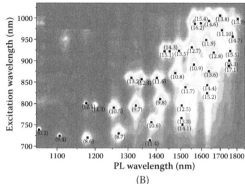

FIGURE 1.9

(A) optical absorption of a purified SWCNT thin film and (B) PL map of SWCNTs identifies their chiralities (Note that nanotubes with $n = m$ or $m = 0$ cannot be detected by PL). Adapted from Optical properties of carbon nanotubes. *Wikipedia*, http://en.wikipedia.org/wiki/Optical_properties_of_carbon_nanotubes (accessed in December 2009) and Kataura, H., et al. 1999. *Synth. Met.* 103: 2555.

resulting from their chiralities, and (2) light trapped in the CNT forests due to multiple reflections. The excellent light absorption property suggests that CNTs can be ideal black bodies widely used in solar energy collectors and infrared thermal detectors[66–69].

PL describes any process in which light is absorbed, generating an excited state, and then light of lower energy is re-emitted upon relaxation to a ground state[62]. The PL of SWCNTs is linearly polarized along the tube axis, thus allowing the CNT orientation to be monitored without direct microscopic observation. In addition, PL of CNTs can offer useful information regarding their structures and physical states as follows:

1. The chiralities (n, m) of SWCNTs to illustrate the presence of semi-conducting SWCNTs in the sample (Figure 1.9B).
2. The physical states of CNTs, whether they occur isolated or in bundle. If CNTs exist in a bundle, there will be no PL due to quenching or shift in energy due to energy transfer processes.
3. Investigation of the effects of solvent, surfactant, and additives on the electrical and optical properties of CNTs.

The PL range of CNTs is rather wide: the emission wavelength can vary between 0.8 and 2.1 μm depending on the CNT structure (Figure 1.9B). However, the PL efficiency of CNTs is usually low, about 0.01%[70], and there is room for enhancement through improving the structural quality of CNTs and clever nanotube isolation strategies. For example, an efficiency of 3% has been reported in CNTs sorted by diameter and length using gradient centrifugation[71], which was further increased to 20% by optimizing the procedure to isolate individual CNTs in a solution[72].

Semiconducting and metallic CNTs can be detected by Raman mapping, and this technique is less sensitive to CNT bundling than photoluminescence. Raman scattering in SWCNTs is resonant, which means that only those CNTs that have one of the band gaps equal to the exciting laser energies are probed[73,74]. The mechanisms behind the Raman characteristics of CNTs will be discussed in Section 1.3.1.3.

The interactions between CNTs, such as chemical linkage, entanglement, and bundling, in general broaden the optical lines. While these interactions strongly affect the PL of CNTs, they have much weaker influences on optical absorption and Raman scattering. Consequently, sample preparation for the latter two characterization techniques is relatively simple.

1.2.6 Defects in CNTs

The importance of defects in CNTs has attracted significant attention in recent years because defects adversely influence the mechanical, electrical, and thermal properties of CNTs. Generally, any heterogeneous atoms and non-hexagonal structures in CNTs can be regarded as defects. They are classified into crystallographic defects and Stone Wales defects[75]. A defect occurring in the form of atomic vacancy is a typical example of crystallographic defects in CNTs. Because of the tiny structure of CNTs, their tensile strengths are dependent on their weakest segments in a manner similar to a chain, where the strength of the weakest link becomes the maximum strength of the chain. High levels of crystallographic defects can lower the tensile strength by up to 85%[75]. Crystallographic defects also affect the electrical properties of CNTs. A common result is a lowered conductivity through the defective region of the tube. A defect in armchair-type CNTs can cause the surrounding region to become semiconducting[61]. Crystallographic defects also cause phonons to scatter, effectively increasing the relaxation rate of the phonons[76], thus reducing both the mean free path and the thermal conductivity of CNTs.

The Stone Wales defects in CNTs create a pentagon and heptagon pair by rearrangement of the C–C bonds. Compared with the hexagonal structure of CNTs, the pentagon and heptagon structures exhibit lower kinetic stability. In other words, the defects originating from these structures are chemically more reactive than the perfect hexagons, thus making them more amenable to the functionalization of CNTs (see Chapter 3 which deals with functionalization of CNTs).

The suitability of CNTs for large-scale applications depends on developing successful strategies to control the defects because the presence of defects often has negative influences on other nanotube properties. Meanwhile, defects in CNTs can bring some benefits, including the introduction of anchor points for chemical functionalization, charge injection, and symmetry-breaking effects, thus facilitating their characterization using various techniques[77]. In addition, defects can cause the tubule to curve. CNTs with different tubular morphologies, such as waved, coiled, and branched, have been fabricated as shown in Figure 1.10[78]. Unlike the usual straight CNTs, these curved CNTs exhibit unique mechanical, electrical, and thermal properties. The utilization of these CNTs with unique morphologies and different properties is expected to have a major impact on the future applications of CNTs.

1.3 Characterization of CNTs

1.3.1 Structural and Morphological Characterization of CNTs

Structural and morphological factors play an important role in determining the properties of CNTs. Therefore, various techniques have been developed and employed to obtain important structural information on the nanoscopic scale. Because the diameters of CNTs are smaller than the wavelengths of visible light (200–600 nm), traditional optical microscopic techniques are not able to characterize the morphology and surface features of CNTs. Therefore, electron microscopic techniques have primarily been employed for this purpose. This section is devoted to reviewing the state-of-the-art techniques employed for structural and morphological characterization of CNTs, and thus providing an essential understanding of the structure–property relations of CNTs and CNT/polymer nanocomposites.

1.3.1.1 Scanning Electron Microscopy (SEM)

SEM employs high-energy electrons to produce images of solid surfaces of length scales down to about 10 nm and provides valuable information regarding the structural arrangement, spatial distribution, and geometrical features of CNTs. The sample must have a moderate electrical conductivity and be stable in a high-vacuum environment. The primary electron beam is scanned over the surface of samples, generating secondary electrons, backscattered electrons, Auger electrons, and x-rays. The Auger electrons and x-rays can be

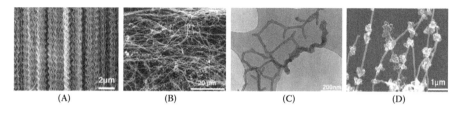

| (A) | (B) | (C) | (D) |

FIGURE 1.10
CNTs with different shapes: (A) waved CNTs, (B) coiled CNTs, (C) branched CNTs, and (D) CNTs with beads. Adapted from Zhang, M., et al. 2009. *Mater. Today* 12: 12.

collected to give the spectroscopic or chemical composition of the scanned materials, whereas the secondary electrons are collected to generate topographical images of the samples[79]. The SEM image shown in Figure 1.11A indicates that CNTs with diameters of about 20 nm and lengths more than 10 μm can be probed, and provides information on the size, size distribution, shapes, physical state such as bundling or entanglement, and so on. SEM can also be employed to examine the distribution of CNTs in the polymer matrix and the fracture morphologies of CNT/polymer nanocomposites, as shown in Figure 1.11B. Large CNT agglomerates are seen from the fracture surface of a CNT/epoxy nanocomposite, suggesting poor dispersion of CNTs. Major failure mechanisms of nanocomposites can be identified by examining the fracture surfaces of nanocomposites taken near the initial crack tip[80,81].

1.3.1.2 Transmission Electron Microscopy (TEM)

TEM uses a much more intense electron beam (200 to 300 keV for most commercial instruments) to generate images. It is a powerful imaging tool to study CNTs at the atomic scale, and usually provides more detailed geometrical features than are seen in SEM images. TEM studies also yield information regarding the number of graphene layers, defects, and diameter of the CNTs, as shown in Figure 1.12A. When operating in the diffraction mode, electron diffraction patterns in a selected area can be made to determine the crystal structure of CNTs.

By coupling the powerful imaging capabilities of TEM with other characterization tools, such as an energy dispersive x-ray (EDX) spectrometer within the TEM equipment, additional properties of the CNTs can be probed. EDX measures the energy and intensity distributions of x-rays generated by the impact of the electron beam on the surface of the sample. The elemental composition within the probed area can be determined to a high degree of precision. This technique is particularly useful for the compositional characterization of functionalized CNTs. Figure 1.12B shows the TEM image of

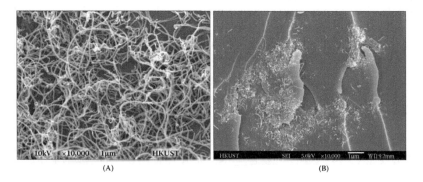

(A)   (B)

FIGURE 1.11
SEM images of (A) dispersed CNTs and (B) CNTs in a polymer matrix. Adapted from Ma, P. C., et al. 2007. *Compos. Sci. Technol.* 67: 2965.

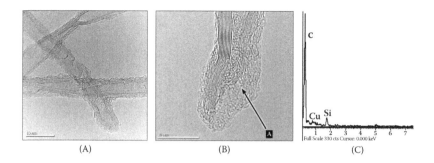

FIGURE 1.12
TEM images of (A) CNTs, (B) silane functionalized CNTs, and (C) corresponding EDX spectrum of silane-functionalized CNTs. Adapted from Ma, P. C., et al. 2006. *Carbon* 44: 3232.

CNT functionalized using an organic silane[82]. Some amorphous materials are attached to the end tips of the MWCNTs (spot A in Figure 1.12B). The detection of silicon by the EDX (Figure 1.12C) confirms that these amorphous materials are derived from the silane molecules. Note that there is a trace amount of copper detected from EDX spectrum; this is coming from the copper grid used for sample preparation.

1.3.1.3 Raman Spectroscopy

Raman spectroscopy is one of the most powerful tools for characterization of CNT structures in a fast and nondestructive manner. This technique provides information about the configuration, diameter, number of walls in CNTs, as well as the presence of crystalline or amorphous carbon. All these capabilities lie in the fact that CNTs are active in the Raman spectrum like other allotropic forms of carbon[83]. However, the position, width, and relative intensity of bands are modified according to the carbon forms[84]. The most distinctive features of Raman spectroscopy for CNTs are shown in Figure 1.13, and the major peaks are assigned as following[79,85]:

1. Radial breathing mode (RBM, inset in Figure 1.13): A series of low-frequency peaks located in the Raman shift range of 100–200 cm^{-1}, a characteristic of SWCNTs assigned to the "breathing" mode of the tubes. The frequency of RBM depends on the diameter of the nanotube.

2. D band ("D" for disorder): A large structure located at around 1360 cm^{-1}, it is assigned to ill-organized graphite or amorphous carbon originating from the structural defects of CNTs.

3. G band ("G" for graphite): This mode corresponds to planar vibrations of carbon atoms and is present in most graphite-like materials between the Raman shift ranges of 1500 and 1600 cm^{-1}. The intensity

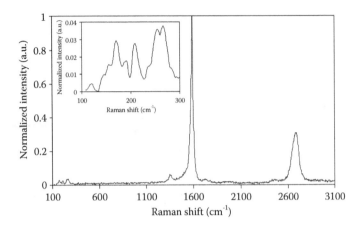

FIGURE 1.13
Raman spectrum of SWCNTs. Inset represents RBM of CNTs.

ratio of the G band to D band, I_G/I_D, is conventionally used to quantify the structural quality of CNTs. The G band in a SWCNT is shifted to lower frequencies relative to graphite (1580 cm^{-1}) and is split into several peaks. The splitting pattern and intensity depend on the tube structure and excitation energy; they can be used, though with much lower accuracy compared to the RBM mode, to estimate the tube diameter and whether the tube is metallic or semiconducting.

4. G′ band: This is a mode observed between 2450 and 2650 cm^{-1} assigned to the first overtone of the defect-induced D band. The position of the G′ mode depends on the CNT diameter, and thus can be used to estimate the SWCNT diameter. In particular, G′ mode is a doublet in double-wall carbon nanotubes, but the doublet is often unresolved due to line broadening.

Other overtones, such as a combination of RBM + G mode at ~1750 cm^{-1}, are frequently seen in CNT Raman spectra. However, they are less important and are not considered here.

The RBM mode is the real signature of the presence of CNTs in a sample because it is not present in graphite. It corresponds to radial expansion–contraction of the nanotube. The RBM frequency, ω, is inversely proportional to nanotube diameter with the relation:

$$\omega \ (\text{cm}^{-1}) = A/\text{diameter (nm)} + B \ (\text{cm}^{-1}) \tag{1.1}$$

where the constants A and B are determined experimentally: A = 223 cm^{-1}/nm, B = 10 cm^{-1}. It was found that the best fits of A and B were not always constant, but the CNTs with a small diameter of 0.8 to 1.3 nm resulted in almost constant values[79].

It should be noted that the Raman spectrum of MWCNTs is influenced by the excitation energy of the laser: the I_G/I_D of MWCNTs were found to decrease at a slope 0.3 eV^{-1} with increasing laser energy[86]. However, the laser energies had only a negligible effect on the intensity ratio of SWCNTs, possibly due to the structural differences between MWCNTs and SWCNTs. This means that the laser energy level should be reported when evaluating the I_G/I_D ratios of MWCNTs.

1.3.2 Characterization of Surface Functionalities on CNTs

1.3.2.1 Infrared (IR) Spectroscopy

IR spectroscopy is an analytical technique that is ideally suited for the characterization of surface functionalities on CNTs, because it allows the measurement of the light absorption in the region of the interband electronic transitions. These electronic transitions are the characteristic signature of CNT electronic structures. There are several IR active modes in SWCNTs depending on their chiralities[85,87]. For MWCNTs, the active modes are observed at about 868 and 1575 cm^{-1} [88,89]. Pristine MWCNTs showed two bands at 3419 and 1058 cm^{-1} [82] (Figure 1.14), which were attributed to the presence of hydroxyl groups (–OH) on their surface, possibly resulting from either atmospheric moisture bound to the MWCNTs or oxidation during purification of the raw material. Another band at 1626 cm^{-1} was assigned to C=O stretching of quinone groups on the surface of MWCNTs.

For CNT/polymer nanocomposites, the introduction of organic functional groups on the CNT surface has been proven to be an effective way to enhance the interfacial interactions between the CNTs and polymer matrix. Numerous studies have been reported that modify the CNT surface properties using organic materials. Table 1.5 summarizes[90,91] the characteristic IR bands of organic functional groups, and is a useful guideline for the interpretation of IR spectra.

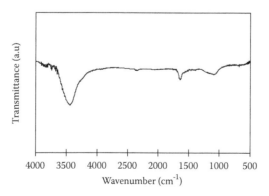

FIGURE 1.14

IR spectrum of pristine MWCNTs (purchased from industry). Adapted from Ma, P. C., et al. 2006. *Carbon* 44: 3232.

TABLE 1.5   Characteristic IR Bands of Organic Functional Groups.

Organic Group	Wavenumber (cm⁻¹)	Assignment	Intensity
Alkane	2853–2962	–C–H stretching	m-s
	1380–1385, 1365–1370	–CH(CH$_3$)$_2$ stretching	s
	1385–1395, ~1365	–C(CH$_3$)$_3$ bending	m-s
Alkene	3010–3095	=C-H stretching	m
	1620–1680	C=C stretching	v
	985–1000, 905–920	R-CH=CH$_2$ stretching	s
	880–900	R$_2$C=CH$_2$ bending	s
	675–730, 960–975	RCH=CHR bending	s
Alkyne	~3300	≡C–H stretching	s
	2100–2260	C≡C stretching	v
Aromatic compound	~3030	Ar-H stretching	v
	690–710, 730–770	Mono-substitution	v, s
	735–770	Di-1, 2- substitution	s
	680–725, 750–810	Di-1, 3- substitution	s
	790–840	Di-1, 4- substitution	s
Alcohol and phenol	3600	–OH (alcohol) stretching	s, b
	3550–3500	–OH (phenol) stretching	s, b
	1300–1000	C–OH stretching	s
Ether	1100	C–O–C stretching	m
Aldehyde and ketone	2700–2900	Aldehyde C–H stretching	m
	1720–1740	Aliphatic aldehyde C=O stretching	s
	1700–1730	Aliphatic ketone C=O stretching	s
	1680–1720	Aromatic aldehyde C=O stretching	m
	1680–1700	Aromatic ketone C=O stretching	m-s
Ester	1730–1750	Aliphatic –C=O stretching	m
	1705–1730	Aromatic –C=O stretching	m-s
	1250–1310	Aromatic –C–O stretching	s
	1100–1300	Aliphatic –C–O stretching	s
Carboxylic acid	2500–3300	–O-H stretching	s, b
	1700	–C=O stretching	s
	1430	–C–O–H in-plane bending	m
	1240	–C–O stretching	m-s
	930	–C–O–H out-of-plane bending	m
Anhydride	1000–1300	–C–O stretching	s
	1740–1780, 1800–1840	–C=O stretching	s

Note: s: strong; m: middle; v: varying; b: broad peak.

Source: Adapted from Tan, Z., et al. 1995. *Experiment textbook for organic chemistry*, 46. Higher Education Press and Stuart, B. 2004. *Infrared spectroscopy: Fundamentals and applications*, 76. New York: Wiley.

1.3.2.2 X-ray Photoelectron Spectroscopy (XPS)

XPS spectra are obtained by irradiating a material with a beam of x-rays while simultaneously measuring the kinetic energy and the number of electrons that escape from the top 1 to 10 nm of the sample. It is a quantitative spectroscopic technique that measures the elemental composition, chemical state, and electronic state of the elements that exist in CNTs. The structural modifications of CNT walls due to chemical interactions with organic compounds or gas adsorption can be studied. Figure 1.15A shows a general XPS spectrum of CNTs with a strong C1s peak at a binding energy of about 284.5 eV. The incorporation of organic silane into CNTs is reflected by the modification of C1s as well as Si2p peaks[82]. The C1s peak presents both a shift and an asymmetric broadening to lower binding energies due to the polar character of the –C–O– and –C–Si– bonds arising from the silane functionalization. The C1s XPS spectrum of CNTs is in general very useful in interpreting the functional groups. It can be simplified into several fitting curves with peaks located at different binding energies, which correspond to features of the functional structures present in CNTs. Table 1.6 summarizes[82] the binding energies and the corresponding assignments from C1s curve fitting, which provides useful information to differentiate the functional groups.

XPS can be employed to study the reaction mechanism between CNTs and organic material. For example, the simplified Si2p signal in Figure 1.15B presents two peaks at different binding energies, indicating two chemical states of Si on the CNT surface: one as bonding between silicon and oxygen originating from MWCNTs, –Si–O–CNT, and the other in the form of siloxane (–Si–O–Si–), resulting from the partial hydrolysis of the silane molecules during the silanization reaction[82,92].

1.3.2.3 Mass Spectrometry (MS)

MS is used for elucidating the fine chemical structures of functional groups attached to CNTs. The technique has both qualitative and quantitative capabilities, including identifying unknown compounds, determining the isotopic composition of elements in a molecule, and determining the structure of a compound by observing its fragmentation. Other applications include quantifying the amount of a compound in a sample or studying the fundamentals of gas phase ion chemistry[93]. One of the unique advantages of this technique is that it can detect the existence of heterogeneous atoms at

TABLE 1.6 Binding Energies and Corresponding Assignments of C1s Curve Fitting.

Binding Energy (eV)	282.8	284.5	285.1–285.3	286.0–286.5	287.6–287.8	289.0–289.4
Assignation	Si–C	$C_g sp^2$	$C_d sp^3$	–C–O	–C=O	–CO–O

Source: Adapted from Ma, P. C., et al. 2006. *Carbon* 44: 3232.

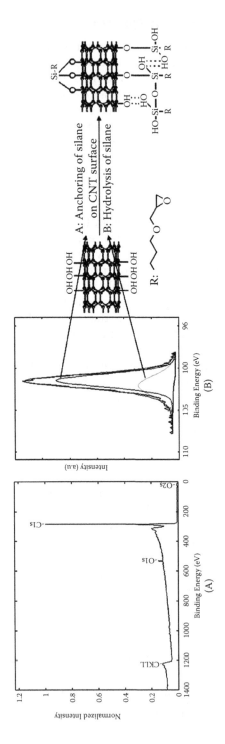

FIGURE 1.15

(A) general XPS spectrum of a pristine CNT, and (B) Si2p XPS spectrum of silane-treated MWCNTs and corresponding reaction mechanism. Adapted from Ma, P. C., et al. 2006. *Carbon* 44: 3232.

a very low concentration in CNTs. Figure 1.16 shows the MS spectrum of amino functionalized CNTs using ball milling (where $N = 0.37$at % as determined by XPS)[94], identifying three different types of nitrogen compounds. They include: (1) the ammonium gas adsorbed onto the CNT surface, as verified by $m/z = 18$; (2) the amines that were verified by the peaks of $-CH_4N$ ($-CH_2NH_2$), $-C_2H_6N$ ($-CH_2CH_2NH_2$), and $-C_3H_8N$ ($-CH_2CH_2CH_2NH_2$) at $m/z = 30, 44$, and 56, respectively; (3) the amide that was present in the form $-C_3H_6NO$ ($-CH_2CH_2-CO-NH_2$) at $m/z = 72$. The comparison between the relative intensities of these nitrogen compounds suggested that only a small amount (~15%) of NH_3 gas was adsorbed onto the CNT surface while the majority of these compounds were covalently bonded to the CNT. Because of its high accuracy, the results from MS are often incorporated as complements to IR and XPS to further verify the functional groups on the CNT surface.

1.3.2.4 Thermogravimetric Analysis (TGA)

TGA is an analytical technique used to determine the thermal stability of a material and its fraction of volatile components by monitoring the change in mass occurring when it is heated. The measurement is normally carried out in air or in an inert atmosphere such as nitrogen and argon, and the mass is recorded as a function of increasing temperature. In the particular case of CNTs, the mass change in an air atmosphere is typically a superposition of the mass loss due to oxidation of carbon into gaseous carbon dioxide and the mass gain due to oxidation of residual metal catalyst into solid oxides[79]. A typical TGA curve of CNTs is illustrated in Figure 1.8A. Although this technique cannot demonstrate the fine chemical structures of functionalized CNTs, it can provide a quantitative analysis of weight changes arising from functional groups on CNTs. In other words, it offers a way to monitor the degree of functionalization of CNTs, information that is particularly important because a severe treatment of CNTs deteriorates their mechanical, electrical, thermal, and other important properties.

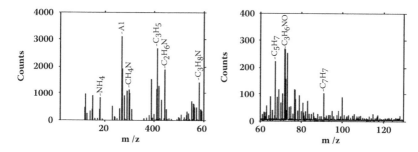

FIGURE 1.16
MS spectra of amino functionalized CNTs using ball milling. Adapted from Ma, P. C., et al. 2008. *Chem. Phys. Lett.* 458: 166.

1.3.3 Summary

The techniques developed to characterize the structural, morphological, and functional properties of CNTs are not limited to those previously mentioned. Only a brief introduction has been provided to help those who are not familiar with them to understand the basic principles and capabilities of the major techniques. Many other techniques, such as scanning probe microscopy (SPM) and PL (discussed in Section 1.2.5) can also be employed for the same purpose. Meanwhile, the characterization of functional groups on CNTs can be accomplished by employing x-ray diffraction (XRD), nuclear magnetic resonance (NMR) spectroscopy, and so on. Table 1.7 summarizes the capabilities of the major techniques available for CNT characterization and the important factors that may affect their selection. It is hoped that the table offers readers a general guideline in selecting proper characterization tools for a specific purpose and end application.

1.4 Other Carbon Nanofillers

1.4.1 Carbon Nanofibers

CNTs and carbon nanofibers (CNFs) are distinguished by the way the cylindrical graphene layers are wrapped. CNTs are perfect cylinders rolled by a single or multiple layers of graphene, whereas CNFs are cylindrical nanostructures with graphene layers arranged as stacked cups or plates, as shown in Figure 1.17[95]. One of the first records of CNFs is probably a patent dated 1889 on synthesis of filamentous carbon material[96]. With the technological development of electronic microscopes, the first observations of CNFs were performed in 1950. Deeper studies of this topic were motivated by the need to inhibit the growth of CNFs because of the persistent problems caused by accumulation of the material in a variety of commercial processes[97]. A more sophisticated understanding of the morphology and properties of CNFs was developed by Endo et al.[98–100].

CNFs have attracted interest because they can provide solutions to some challenging problems in fiber-reinforced composites. With regard to physical dimensions, mechanical and structural performance, and product cost, CNFs complete a continuum bounded by nanoscopic fillers like CB, fullerene, grapheme, and CNTs on one end, and continuous carbon fibers on the other, as illustrated in Table 1.8. CNFs are able to combine many advantages of other forms of carbon materials for reinforcement of engineering polymers for high mechanical performance. They have better transport and mechanical properties than carbon fibers as shown in Table 1.9, and can be produced in high volumes at a low cost. In equivalent production volumes, CNFs are projected to have a cost comparable to E-glass fibers on a per-pound basis,

TABLE 1.7 Summary of the Techniques Used for CNT Characterization.

Technique / Capability/Factor	TEM	SEM	SPM	Raman	IR	PL	XPS	XRD	MS	TGA	NMR
CNT chirality	×	×	×	√	×	√	×	√	×	×	×
Dimension of CNTs	√	√	√	√	×	×	×	√	×	×	×
Crystal lattice of CNTs	√	×	√	×	×	×	×	√	×	×	×
Defect in CNTs	√	×	√	√	√	×	√	√	×	√	×
Heterogeneous atom on CNTs	√	×	√	√	×	×	√	√	√	√	√
Functional group on CNTs	×	×	×	√	√	×	×	√	√	√	√
Damage to CNTs	√	√	×	×	×	×	√	√	√	√	×
Quantitative analysis	×	×	×	√	×	×	√	√	×	√	×
Amount of sample required	<mg	<mg	<mg	~mg	~mg	<mg	~mg	>mg	~mg	~mg	<mg
Form of results	Image [a]	Image [a]	Image	Spectrum	Spectrum	Image	Spectrum[b]	Spectrum	Spectrum	Spectrum	Spectrum
Data interpretation	Easy	Easy	Easy	Easy	Not easy	Not easy	Not easy	Not easy	Not easy	Easy	Not easy
Sample preparation [c]	Dis	Au	Dis	No	KBr	Dis	No	No	No	No	Dis
Cost [d]	High	Middle	Low	Middle	Low	Low	High	High	High	Low	High

Notes: [a] Spectrum can also be obtained with EDX mode. [b] Elemental composition of materials can be expressed as a table. [c] Acronyms: Dis: Dispersion of CNTs in an organic solvent is needed; Au: Gold coating on sample to enhance conductivity; No: No need for sample preparation; KBr: CNTs are mixed with potassium bromide (KBr) and pressed into a pellet for characterization (most cases). [d] Refers to the cost on the operation and maintenance of equipment.

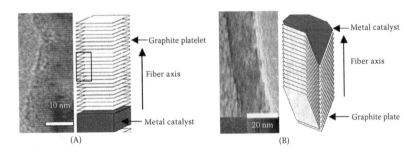

FIGURE 1.17
High-resolution electron micrographs and schematic illustrations of CNFs with different arrangements of graphene layers: (A) plates and (B) stacked cups. Adapted from Terry, R., et al. http://ftp.wtec.loyola.edu/loyola/nano/US.Review/09_03.htm, accessed in November 2010 and Kim, Y. A., et al. 2005. *Carbon* 43: 3005.

yet they possess much better mechanical, electrical, and thermal properties than glass and are equal to, or exceed, much more costly commercial carbon fibers[101].

CNFs can be produced by several different methods, similar to CNT production. Catalytic CVD is commonly used, which involves catalytic decomposition of hydrocarbon gases or carbon monoxide on selected metal particles like iron, cobalt, nickel, and some of their alloys. In contrast to the CVD method used to grow CNTs, however, the growth of CNFs requires a low temperature of 500–700°C. Other techniques, including the floating cata-lyst method, co-catalyst deoxidization, polymer blend techniques, and the template-assisted method have been developed for CNF synthesis in recent years[101]. It is now possible to produce CNFs in different grades with different wall thicknesses. Most significantly, commercial quantities of CNFs can now be produced, followed by surface functionalization with carboxylic groups, making them suitable for further functionalization or applications in areas like biomedical sensors and drug delivery[102].

It should be noted that regardless of the techniques employed for CNF fabrication, the as-produced CNFs contain a discontinuous structure of graphene planes. Therefore, a thermal treatment is necessary to improve the crystallinity of the carbon structure[103], and thus the mechanical and electrical properties of CNFs. The perfect rolling of graphene layers results in the formation of CNTs, which exhibit much better properties than CNFs. However, the high cost and poor processing of CNTs have hampered efforts to exploit the anticipated composite properties. As much ensuing work evi-denced, it is clear that the van der Waals interactions and the high aspect ratios of CNTs are a hindrance for dispersion and the following composite preparation. It is much more likely that CNFs will be more popularly uti-lized than CNTs in applications requiring less demanding manufacturing processes.

TABLE 1.8 Comparison of Dimensions and Costs of Different Carbon Materials.

Carbon Materials	Description	Dimension	Cost (US$/g)
Graphene		<1 nm in thickness, length in μm scale	N/A
Fullerene (C_{60}, C_{70}, etc)		0.7–2 nm in diameter	40–70
SWCNT		1–10 nm in diameter, various length	20–100
MWCNT		10–50 nm in diameter, various length	0.5–5
CNFs		50–200 nm, 100 μm in length	0.05–0.2
Carbon fiber		5–20 μm in diameter, various length	0.02–0.1
CB		20–100 nm in diameter	<0.02
Graphite		10–1000 μm in diameter, various length	<0.002

TABLE 1.9 Physical Properties of CNFs and Carbon Fibers.

Physical Properties	C Filament	CNFs	Carbon Fiber
Density (g/cm³)		1.8	1.8
Tensile modulus (GPa)		400–600	200–350
Tensile strength (GPa)		5–7	2–5
Electrical conductivity (S/cm)		500–1000	10–100
Thermal conductivity (W/(m·K))		20–100	1–10

Source: Adapted from Kang, I., et al. 2006. *Compos B* 37: 382, and Kroto, H. W., et al. 1985. *Nature* 318: 162.

1.4.2 Spherical Carbonaceous Nanofillers

Spherical carbonaceous nanofillers have been used for many years in the plastics industries. The use of spherical carbon nanofillers, such as CB, dates back many centuries when the Chinese used "lamp black" as a pigment in black ink in the third century B.C. With the development of science and technology, new forms of spherical carbonaceous nanofillers, such as nanodiamond, fullerene, and onion-like carbon, are observed, synthesized, and employed as fillers for nanocomposites. There is extensive literature describing the fabrication processes, properties, and applications of these nanofillers. Table 1.10 briefly summarizes their structures, fabrication methods, and applications.

TABLE 1.10 Spherical Carbonaceous Nanofillers.

Nanofillers	History	Structure	Fabrication Process	Major/Potential Applications
CB	"Lamp black" in the 3rd century B.C. Scale production in the 1970s	Spherical particles of 10–60 nm coalesced into chain-like agglomerates	Incomplete furnace combustion of tar and vegetable oil (small amount)	Xerographic toner, UV protection; Conducting filler; reinforcement fillers for rubbers (90% of CB is used in tire industry)
Nanodiamond	In 1963, classified research in Soviet Union revealed its production	Isometric-hexoctahedral crystal lattice. 2–20 nm in diameter with average particle sizes of 4–5 nm	Detonation of explosive; Suspension of graphite in organic liquid under ultrasonication	Lapping and polishing; Additives to engine oils; Reinforcing fillers for plastics and rubbers; Dry lubricants; Additives to galvanic electrolytes
Fullerene	Discovery of C_{60} in 1985, and shortly thereafter came to the discovery of other fullerenes	<1.0 nm in diameter with 60 or more C atoms bonded into hexagonal and pentagonal shapes like a soccer ball	Arc discharge of graphite; Combustion of benzene; Condensation of aromatic hydrocarbon	Plastic solar cell; Biological applications (carrier for some biomolecules); Battery electrodes; Functional thin film; Sensor; Field emission display, etc.
Onion-like carbon	First observation from graphitized CB using higher resolution TEM in 1980[104]	Quasi-spherical nanoparticles consisting of fullerene-like C layers enclosed by concentric graphene layers	Electron irradiation of graphite; Thermal annealing of nanodiamond at 1000–1800 K	Magnetic recording systems; Magnetic fluids; Electromagnetic shielding materials; Nanocapsules for drug delivery systems; Optical limiter

1.5 Other Nanotubes

The stability cylindrical material forms is not limited to CNTs. In 1992, scientists discovered a new kind of nanotube composed of tungsten and sulfur (Figure 1.18A)[105]; thus began the subject of inorganic nanotubes (INTs). Theoretical simulations showed that any of the elements and compounds known to form stable sheet or layered structures can be rolled into tubes[106]. Several methods have been developed to fabricate INTs, as summarized in Table 1.11[107,108]. Each method may result in an INT with a specific atomic structure, chirality, purity, diameter, and length. A common method is the

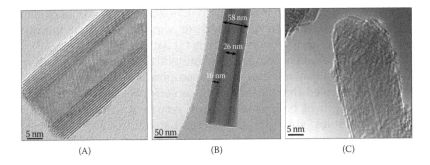

(A)　　　　　　　(B)　　　　　　　(C)

FIGURE 1.18
TEM images of different INTs: (A) WS_2 NT, (B) halloysite NT, and (C) boron nitride NT. Adapted from Tenne, R., et al. 1992. *Nature* 360: 444; Ye, Y. P., et al. 2007. *Polymer* 48: 6426; and Wang, J., et al. 2005. *Nano. Lett.* 5: 2528.

TABLE 1.11　Summary of Various INTs and Their Fabrication Methods.

Item INTs	Method	Synthetic Route
WS_2, MoS_2, ZnS	Sulfurization	Heating MoO_x or WO_x in H_2S; Heating ZnO in H_2S
NbS_2, TaS_2, HfS_2, ZrS_2, Vox	Decomposition of precursor crystals	Oxidation of tri- and tetra-transition metal chalcogenides at elevated temperatures
TiO_2, Al_2O_3, GaN, Au, Co, Fe, Si, Ag	Template growth	Sol-gel method, electrodeposition
VO_x	Layered precursors	Solvothermal synthesis
Bi	Lameallar precursors	Hydrothermal pyrolysis
InGaAs/GaAs	Misfit rolling	Rolling up of strained heterostructures
MoS_2, Au-MoS_2, WS_2, $PbNb_nS_{2n+1}$	Direct synthesis from vapor phase	Chemical transport reaction

Source: Remskar, M. 2004. *Adv. Mater.* 16: 1497 and Tenne, R., et al. 2009. *Annu. Rev. Mater. Res.* 39: 2.1.

exposure of precursors to high temperatures by laser heating or an arc discharge. Another technique is the direct synthesis of INTs from the vapor phase via chemical transport reactions. Template-assisted synthesis is perhaps the most promising method, since it allows more precise control of the nanotube type and dimensions.

INTs are heavier than CNTs and not as strong under tensile stress, but they are particularly strong under compression, indicating potential applications in impact-resistant structures such as bulletproof vests[105]. As the diversity of available nanotubes increases, INTs can be obtained with a wide variety of molecular shapes, such as columns, pipes, bearings, and springs. The mechanical properties of these tubes are controllable by electronic means, making them ideal components for nano-electro-mechanical systems (NEMS). Some INTs, consisting of transition metals and oxides, such as vanadium oxide and manganese oxide, have been developed for various catalytic applications, gas-sensing devices, and cathode materials for batteries[107–110].

INTs are also observed in some natural mineral deposits, such as halloysite (Figure 1.18B). It is a clay mineral with the empirical formula $Al_2Si_2O_5(OH)_4$. Recent studies showed that halloysite was an effective modifier to increase the impact strength of halloysite/polymer composites without scarifying the flexural modulus, strength, and thermal stability[109]. INTs constructed from the main group elements of boron and nitrogen were also developed (Figure 1.18C). Being an isoelectronic material with benzene, the boron nitride nanotubes (BNNTs) possess uniform electronic properties that are insensitive to their diameters and chiralities. In addition, BNNTs demonstrated high oxidation resistance at temperatures up to 800°C, excellent piezoelectricity, and a potential application for room-temperature hydrogen storage. Superlattices or isolated CNT–BNNT junctions are also known to produce itinerant ferromagnetism and spin polarization, making BNNTs ideally suited for innovative applications in various branches of science[110].

References

1 Donnet, J. B., et al. 1993. *Carbon black: Science and technology*, 2nd ed. New York: Marcel Dekker, Inc., 1–5.

2 Carbon. *Wikipedia*, http://en.wikipedia.org/wiki/Carbon (accessed November 2009).

3 Biological abundance of elements. *Internet encyclopedia of science*, http://www.daviddarling.info/encyclopedia/E/elbio.html (accessed in November 2009).

4 Kang, I., et al. 2006. Compos B 37: 382.

5 Ma, P.C., et al. 2010. Compos A 41: 1345.

6 Dresselhaus, G. 2001. *Carbon nanotubes: Synthesis, structure, properties and applications*, 77–80. New York: Springer.

7 Monthioux, M., et al. 2006. *Carbon* 44: 1621.

8 Fiorito, S. 2008. *Carbon nanotubes: Angels or demons*, 1–2. Singapore: Pan Stanford.

9 Oberlin, A., et al. 1976. *J. Cryst. Growth* 32: 335.

10 Kroto, H. W., et al. 1985. *Nature* 318: 162.

11 Krätschmer, W., et al. 1990. *Nature* 347: 354.

12 Iijima, S. 1991. *Nature* 354: 56.

13 Iijima, S. 1993. *Nature* 363: 603.

14 Bethune, D. S., et al. 1993. *Nature* 1993 363: 605.

15 Ebbesen, T. W., et al. 1993. *Chem. Phys. Lett.* 209: 83.

16 Thostenson, E. T., et al. 2001. *Compos. Sci. Technol.* 61: 1899.

17 Hu, Y., et al. 2006. *Rep. Prog. Phys.* 69: 1847.

18 Chen, Y., et al. 2006. *Sci. Technol. Adv. Mater.* 7: 839.

19 Wilson, M. 2002. *Nanotechnology: Basic science and emerging technologies*, 100. Boca Raton, FL: Chapman & Hall/CRC.

20 Reich, S., et al. 2004. *Carbon nanotubes: Basic concepts and physical properties*, 3. New York: Wiley-VCH.

21 Dresselhaus, M. S., et al. 1995. *Carbon* 33: 883.

22 Salvetat-Delmotte, J. P., et al. 2002. *Carbon* 40: 1729.

23 Yakobson, B. I., et al. 2001. *Topics Appl. Phys.* 80: 287.

24 Sinnott, S. B., et al. 1998. *Carbon* 36: 1.

25 Gao, G. H., et al. 1998. *Nanotechnology* 9: 184

26 Hernández, E., et al. 1998. *Phys. Rev. Lett.* 80: 4502

27 Treacy, M., et al. 1996. *Nature* 381: 678.

28 Yu, M. F., et al. 2000. *Science* 287: 637.

29 Terrones, M. 2003. *Annu. Rev. Mater. Res.* 33: 419.

30 Popov, V. N. 2004. *Mater. Sci. Engin. R.* 43: 61.

31 Yamamoto, G., et al. 2005. *J. Jpn. Soc. Powder. Metal.* 52: 826.

32 Coleman, J. N., et al. 2006. *Carbon* 44: 1624.

33 Salvetat, J. P., et al. 1999. *Phys. Rev. Lett.* 82: 944.

34 Yu, M. F., et al. 2000. *Phys. Rev. Lett.* 84: 5552.

35 Ajayan, P.M. 1999. *Chem. Rev.* 99: 1787.

36 Saito, R., et al. 1992. *Phys. Rev. B* 46: 1804.

37 Hong, S., et al. 2007. *Nature Nanotechnology* 2: 207.

38 Tans, S.J., et al. 1997. *Nature* 386: 474.

39 Kim, T. H., et al. 2008. *Nanotechnology* 19: 485201.

40 Dai, H., et al. 1996. *Science* 272: 523.

41 Ebbesen, T. W., et al. 1996. *Nature* 382: 54.

42 Saito, R., et al. 1996. *Phys. Rev. B* 53: 2044.

43 Langer, L., et al. 1996. *Phys. Rev. Lett.* 76: 479.

44 Nakayama, Y., et al. 1995. *Jpn. J. Appl. Phys.* 34: 10.

45 De Heer, W. A., et al. 1995. *Science* 270: 1179.

46 Terrones, M., et al. 1998. *Appl. Phys. A* 66: 307.

47 Baumgartner, G., et al. 1997. *Phys. Rev. B* 55: 6704.

48 Berber, S., et al. 2000. *Phys. Rev. Lett.* 84: 4613.

49 Buerschaper, R. A. 1944. *J. Appl. Phys.* 15: 452.

50 Yi, W., et al. 1999. *Phys. Rev. B* 59: R9015.

51 Kim, P., et al. 2001. *Phys. Rev. Lett.* 87: 215502.

52 Hone, J., et al. 2000. *Appl. Phys. Lett.* 77: 666.

53 Thess, A., et al. 1996. *Science* 273: 483.

54 Hone, J., et al. 1999. *Phys. Rev. B* 57: R2514.

55 Hone, J., et al. 2000. *Science* 289: 1730.

56 Pop, E., et al. 2006. *Nano. Lett.* 6: 96.
57 Zhang, H. L., et al. 2005. *J. Appl. Phys. Lett.* 97: 1143101.
58 Chiu, H. Y., et al. 2005. *Phys. Rev. Lett.* 95: 226101.
59 Pang, L. S., et al. 1993. *J. Phys. Chem.* 97: 6941.
60 Saxby, J. D., et al. 1992. *J. Phys. Chem.* 96: 17.
61 Thostenson, E., et al. 2005. *Compos. Sci. Technol.* 65: 491.
62 Optical properties of carbon nanotubes. *Wikipedia*, http://en.wikipedia.org/wiki/Optical_properties_of_carbon_nanotubes (accessed in December 2009).
63 Sinnott, S. B., et al. 2001. *Critical Rev. Solid State Mater. Sci.* 36: 145.
64 Kataura, H., et al. 1999. *Synth. Met.* 103: 2555.
65 Mizuno, K., et al. 2009. *Proceedings of the National Academy of Sciences* 106: 6044.
66 Nunes, C., et al. 2002. *Vacuum* 67: 623.
67 Cao, A., et al. 2002. *Sol. Energy Mater. Sol. Cells* 70: 481.
68 Lira-Cantu, M., et al. 2005. *Sol. Energy Mater. Sol. Cells* 87: 685.
69 Shashikala, A. R., et al. 2007. *Sol. Energy Mater. Sol. Cells* 91: 629.
70 Wang, F., et al. 2004. *Phys. Rev. Lett.* 92: 177401.
71 Crochet, J., et al. 2007. *J. Am. Chem. Soc.* 129: 8058.
72 Ju, S. Y., et al. 2009. *Science* 323: 1319.
73 Fantini, C., et al. 2004. *Phys. Rev. Lett.* 93: 147406.
74 Souza Filho, A. G., et al. 2004. *Phys. Rev. B* 69: 115428.
75 Carbon nanotube. *Wikipedia*, http://en.wikipedia.org/wiki/Carbon_nanotube (accessed in November 2009).
76 Mingo, N., et al. 2008. *Phys. Rev. B* 77: 033418.
77 Jorio, A., et al. 2008. *Carbon nanotubes: Advanced topics in the synthesis, structure, properties and applications*, 5. New York: Springer.
78 Zhang, M., et al. 2009. *Mater. Today* 12: 12.
79 Freiman, S., et al. 2008. *Measurement issues in single wall carbon nanotubes*, 960–19. National Institute of Standards and Technology (NIST) Special Publication. Washington, DC: U.S. Commerce Department.
80 Ma, P. C., et al. 2007. *Compos. Sci. Technol.* 67: 2965.
81 Ma, P. C., et al. 2009. *ACS Appl. Mater. Interfaces* 1: 1090.
82 Ma, P. C., et al. 2006. *Carbon* 44: 3232.
83 Arepalli, S., et al. 2004. *Carbon* 42: 1783.
84 Ferrari, A., et al. 2000. *Phys. Rev. B* 61: 14095.
85 Belin, T., et al. 2005. *Mater. Sci. Engin. B* 119: 105.
86 Ouyang, Y., et al. 2008. *Physica E* 40: 2386.
87 Kuhlmann, K., et al. 1998. *Chem. Phys. Lett.* 294: 237.
88 Kastner, K., et al. 1994. *Chem. Phys. Lett.* 221: 53.
89 Eklund, P., et al. 1995. *Carbon* 33: 959.
90 Tan, Z., et al. 1995. *Experiment textbook for organic chemistry*, 46. Beijing: Higher Education Press.
91 Stuart, B. 2004. *Infrared spectroscopy: Fundamentals and applications*, 76. New York: Wiley.
92 Velasco-Santos, C., et al. 2002. *Nanotechnology* 13: 495.
93 Mass spectrometry. *Wikipedia*, http://en.wikipedia.org/wiki/Mass_spectrometry (accessed in November 2009).
94 Ma, P. C., et al. 2008. *Chem. Phys. Lett.* 458: 166.
95 Terry, R., et al. http://ftp.wtec.loyola.edu/loyola/nano/US.Review/09_03.htm, accessed in Nov, 2010.

 96 Hughes, T. V., et al. U.S. Patent No. 405, 480, 1889.
 97 Schlogl, R., et al. U.S. Patent No. 20090220767, 2009.
 98 Endo, M., et al.1992. *J. Phys. Chem. Solids* 96: 6941.
 99 Endo, M., et al. 2003. *Carbon* 41: 1941.
100 Kim, Y. A., et al. 2005. *Carbon* 43: 3005.
101 Advani, S. G. 2007. Processing and properties of nanocomposites, 141–150. New Jersey: WSP.
102 Nadarajah, A., et al. 2008. *Key Engin. Mater.* 380: 193.
103 Tibbetts, G. G., et al. 2007. *Compos. Sci. Technol.* 67: 1709.
104 Iijima, S. 1980. *J. Cryst. Growth* 50: 675.
105 Tenne, R., et al. 1992. *Nature* 360: 444.
106 Enyashin, A. N., et al. 2006. *Springer Series Mater. Sci.* 93: 33.
107 Remskar, M. 2004. *Adv. Mater.* 16: 1497.
108 Tenne, R., et al. 2009. *Annu. Rev. Mater. Res.* 39: 2.1.
109 Ye, Y. P., et al. 2007. *Polymer* 48: 6426.
110 Wang, J., et al. 2005. *Nano. Lett.* 5: 2528.

2

Dispersion of CNTs

2.1 Introduction

Many research efforts have been directed toward producing carbon nanotubes (CNT)/polymer nanocomposites for functional and structural applications[1-5]. However, even after a decade of research, the full potential of employing CNTs as reinforcements is severely limited because of the difficulties associated with dispersion of CNTs during processing and poor interfacial interaction between the CNTs and the polymer matrix. The nature of the dispersion problem for CNTs is different from other conventional fillers, such as spherical particles and carbon fibers, because CNTs have a diameter on a nanometer scale with a high aspect ratio (>1000) and an extremely large surface area. In addition, commercial CNTs are supplied in the form of heavily entangled bundles, aggravating the difficulties of dispersion in a polymer. In this chapter, fundamental issues regarding CNT dispersion, including the nature of dispersion problems, techniques and mechanisms for dispersion, and dispersion behavior of CNTs affected by solvents and surfactants are discussed.

2.2 Dispersion Behavior of CNTs

2.2.1 Dispersion and Distribution of CNTs

One of the key limitations of CNTs for practical applications is their lack of dispersability in other media. Without proper dispersion or distribution, the excellent properties of CNTs cannot be fully exploited. The dispersion of CNTs in solvents and polymer matrices has been studied extensively in recent years[3-5]. These studies on the morphologies of CNTs taken at different magnifications suggest that the degree of CNT dispersion in a polymer matrix can be totally different depending on the techniques and scales used for characterization. It should be noted that previous studies often confusingly used the terms dispersion and distribution; thus, it is important to clearly define these two terms and identify any differences before there is any related discussion. For conventional composites, *distribution* describes the homogeneity of fillers throughout the material, whereas the term *dispersion* is defined as an "even distribution" of fine particles in a matrix medium, which describes the level of agglomeration[6]. In contrast, for CNTs and CNT/polymer nanocomposites, dispersion has two aspects: (1) *disentanglement* of CNT bundles or agglomerates, which is a *nanoscopic dispersion*, and (2) *uniform distribution* of individual CNTs throughout the medium, which is more of a *micro-* and *macroscopic dispersion*. Figure 2.1 schematically illustrates four distinct states of CNTs in a medium: (A) good distribution but poor dispersion, (B) poor distribution and poor dispersion, (C) poor distribution but good dispersion, and (D) good distribution and good dispersion.

2.2.2 Nature of Dispersion Problems for CNTs

The nature of problems for CNT dispersion is different from other conventional fillers because of the extremely large surface area created by CNTs in a polymer matrix. Table 2.1 compares the dimensions of commonly used fillers and the number of filler particles in a composite corresponding to a constant volume fraction of 0.1%[7]. These data may shed light on why the dispersion of CNTs in a polymer matrix is more difficult than other fillers. It can be seen that the dimensions of fillers have a significant effect on the number of particles in a given volume of composites: While there are only about 2 particles for Al_2O_3 spheres, the number increases to 200 for carbon fibers, and 442 million for CNTs, thus indicating increasingly more difficult dispersion of the nanoparticles with a high aspect ratio than the microscopic spherical fillers.

The three-dimensional distribution of these micro- and nanoscale fillers in a polymer matrix is schematically shown in Figure 2.2, which gives a vivid depiction of different distribution behaviors affected by the geometry and size of fillers. The distribution of microscale fillers (A and B in Figure 2.2) is homogeneous throughout the matrix, and the individual particles are clearly seen with a large volume of matrix-rich region at the low filler volume fraction.

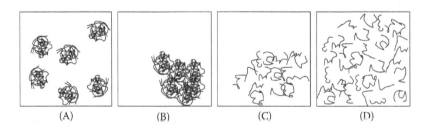

(A)   (B)   (C)   (D)

FIGURE 2.1
A schematic illustration showing the differences between dispersion and distribution states of CNTs: (A) good distribution but poor dispersion, (B) poor distribution and poor dispersion, (C) poor distribution but good dispersion, and (D) good distribution and good dispersion.

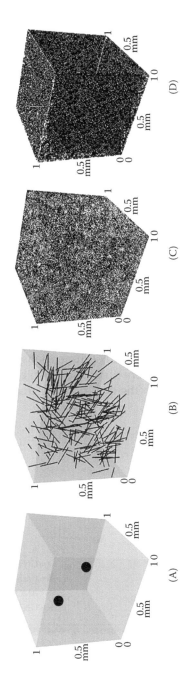

FIGURE 2.2

Distribution of micro- and nanoscale fillers with a fraction of 0.1 vol% in a reference cube of 1 mm^3 in volume (A: Al$_2$O$_3$ particle; B: Carbon fiber; C: GNP; D: CNT).

In contrast, when graphite nanoplatelets (GNPs) and CNTs of the same filler volume fraction are filled, the whole volume is covered by the particles (C and D in Figure 2.2). Considering the particle agglomeration due to the electrostatic interaction and van der Waals force, the real distribution of nanoscale fillers should be more complicated than the schematics shown here.

2.2.3 Surface Area and Interactions between CNTs

The relatively large quantity of particles and their size effect lead to an exceptionally large surface area of CNTs in a composite. Figure 2.3 shows the total surface areas of different types of fillers calculated using the equations given in Table 2.1 under the assumption that the fillers are dispersed uniformly in the composites. For the fillers with size less than 100 nm, such as CNTs and GNPs, the surface areas are a few orders of magnitude larger than their counterparts with sizes in micrometer scale, namely Al_2O_3 particles and carbon fibers. The larger surface area means a larger interface or interphase volume created between the fillers and matrix material. A classic definition of the interface in a composite is a surface formed by a common boundary of reinforcing fillers and matrix that is in contact and maintains the bond in between for load transfer[8]. It is a region with altered chemistry, altered polymer chain mobility, altered degree of cure, and altered crystallinity that is unique from those of the fillers or the matrix. The thickness of interface in CNT composites has been reported to be as small as 2 nm and as large as about 500 nm depending on the CNT dimensions[1,8]. Even if the interfacial region is only a few nanometers thick, the polymer matrix exhibits a behavior completely different from the neat polymer due to the aforemen-

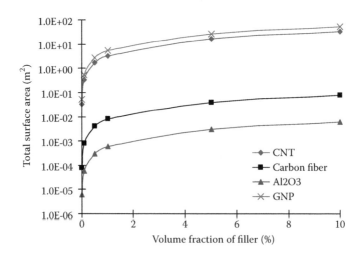

FIGURE 2.3
Total surface area of fillers in composites with varying filler volume fractions.

TABLE 2.1   Dimension and Corresponding Number of Particles in Composites for Different Fillers.

Filler \ Description	Average Dimension of Filler	Density (g/cm³)	N*	S**
Al₂O₃ particle	100 μm in diameter (d)	4.0	1.9	$S = \pi d^2$
Carbon fiber	5 μm in diameter (d) × 200 μm in length (l)	2.25	255	$S = \pi dl + \pi d^2/2$
GNP	45 μm in length (square, l), 7.5 nm in thickness (t)	2.2	$6.58{\times}10^4$	$S = 4l^2 + 2lt$
CNT	12 nm in diameter (d) × 20 μm in length (l)	1.8	$4.42{\times}10^8$	$S = \pi dl + \pi d^2/2$

Notes: * N: Number of particles in 1.0 mm³ with 0.1 vol% filler content; ** S: Surface area of individual particles.

tioned changes that all adversely affect the viscosity and thus the dispersion of CNTs in the polymer matrix.

In addition to the size effect of CNTs, their physical nature also plays an important role in dispersing them into a polymer matrix. As-produced CNTs are held together either in entanglements (for multiwalled CNTs [MWCNTs]) or bundles consisting of 50 to a few hundred individual CNTs (for single-walled CNTs [SWCNTs]) due to van der Waals forces, as illustrated in Figure 2.4. There is ample evidence showing that these bundles and agglomerates often result in much lower mechanical and electrical properties of composites than those by theoretical predictions based on well-dispersed individual CNTs[1-5]. It is a real challenge to separate or disentangle these bundles and agglomerates into individual CNTs without damaging the inherent structure or fragmenting them into smaller pieces.

(A)   (B)

FIGURE 2.4
Electronic microscope images of as-produced CNTs: (A) TEM image of SWCNT bundle, adapted from Ma, P.C, 2008. PhD thesis. Novel surface treatment, functionalization and hybridization of carbon nanotubes and their polymer-based composites. Hong Kong University of Science and Technology, 8–10. (B) SEM image of entangled MWCNT agglomerates.

Therefore, the dispersion of CNTs is not only a geometrical problem dealing with the length and size of CNTs, but also depends on the techniques used to separate the CNT agglomerates into individual CNTs and stabilize them in a polymer matrix to avoid secondary agglomeration.

2.3 Techniques for Mechanical CNT Dispersion

Section 2.2 clearly indicates that the incorporation of CNTs into a polymer matrix inevitably creates an extremely large interface area or volume due to the large surface area of fillers, exacerbating the dispersion problem. There is a sizable volume of literature reported on the techniques developed for CNT dispersion in polymer matrices[10–15]. However, little information has hitherto been reported on the principles and characteristics of these dispersion techniques. Basically, there are two distinct approaches to CNT dispersion: (1) purely mechanical methods and (2) combinations of functionalization and mechanical dispersion. This section presents the principles of the first type of mechanical dispersion method along with some typical results obtained by employing these methods. A general guideline in selecting proper techniques for CNT dispersion is also presented. The methods based on CNT functionalization and mechanical dispersion will be discussed in Chapter 3.

2.3.1 Ultrasonication

Ultrasonication is one of the most frequently used methods for nanoparticle dispersion. *Ultrasonication* is the act of applying ultrasound energy to agitate particles usually using an ultrasonic bath or an ultrasonic probe or horn (A and B in Figure 2.5), also known as a *sonicator*. When ultrasound propagates through a liquid medium via a series of compressions, attenuated

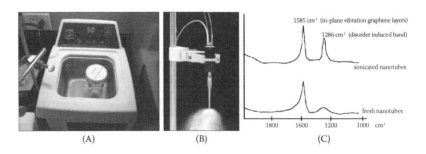

(A)   (B)   (C)

FIGURE 2.5
Sonicators with different modes for CNT dispersion: (A) water bath sonicator, (B) probe/horn sonicator), (C) effect of sonication on the structure of CNTs: Raman spectra of CNTs before and after sonication. Adapted from Lu, K. L., et al. 1996. *Carbon* 34: 814.

waves are induced in the molecules of the medium through which it passes. The production of these shock waves promotes the "peeling off" of individual nanoparticles located at the outer part of the nanoparticle bundles or agglomerates, resulting in separation of individualized nanoparticles from the bundles[16]. Ultrasonication is an effective method to disperse CNTs in some low-viscosity liquids, such as water, acetone, ethanol, and so on. Because most polymers are either in solid or viscous states, they are often dissolved in a solvent before sonication treatment of CNTs.

Standard laboratory sonicators in the form of a water bath run at 20–23 kHz with a power less than 100 watts. Commercial probe sonicators have adjustable amplitude and power ranging from 20–70% and 100–1500 W, respectively. The probe is usually made of an inert metal such as titanium. Most probes are attached with a base unit and then tapered down to a point with diameter of from 1.6 to 12.7 mm[16], allowing the energy to be concentrated on the tip, thus giving the probe high intensities. One consequence of this configuration is that sonication can rapidly generate substantial heat. Therefore, if CNTs are sonicated in volatile solvents, such as ethanol and acetone, the solvent bath must be kept cold—for example, by placing the bath in ice—and the sonication must be carried out in short intervals.

If the sonication treatment is excessive or too long, it can lead to a rupture and damage to CNTs, especially when the probe sonicator is employed. It has been confirmed that prolonged ultrasonication can result in a significant increase of the Raman D-band (Figure 2.5C), which represents disordered carbon materials on CNTs, suggesting the generation of defects on the CNT surface[17]. In some cases, the graphene layers of CNTs were destroyed and the nanotubes were converted into amorphous carbon nanofibers[18]. Localized damage of CNTs deteriorated the electrical and mechanical properties of the CNTs and the corresponding CNT/polymer nanocomposites.

2.3.2 Calendering

Calendering, which is commonly known as a *three–roll mill*, is a machine tool that employs the shear force created by rollers to mix, disperse, or homogenize viscous materials. This method has been designed to disperse color pigments for cosmetics and lacquers. The general configuration of a calendering machine consists of three adjacent cylindrical rollers that run in different directions (Figure 2.6A)[19–21]. The first and third rollers, called the *feeding* and *apron* rollers (*n1* and *n3* in Figure 2.6B), rotate in the same direction while the central roller rotates in the opposite direction. A pasty substance containing nanofillers is fed into the hopper, where it is drawn between the feed and central rollers. When predispersed, the substance sticks to the bottom of the central roller, which transports it into the second gap. In this gap, the paste is dispersed to the desired degree of fineness. Upon exiting, the material that remains on the central roller moves through the second nip between the central roller and apron roller, subjecting the paste to an even higher shear force due to the

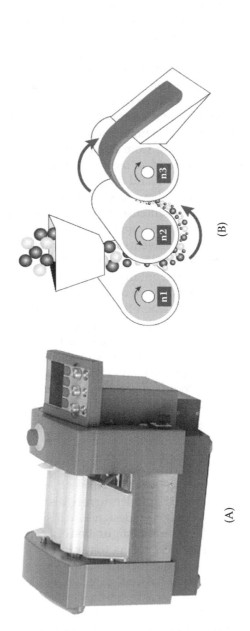

FIGURE 2.6
(A) Calendering or three–roll mill machine used for particle dispersion into a polymer matrix, and (B) the general configuration and working mechanism. Adapted from EXAKT. Three roll mills, http://www.exakt.com/three-roll-mills.25tM52087573abo.o.html (accessed June 2010).

higher speed of the apron roller. A knife blade then scrapes the processed material off the apron roller and transfers it to the apron. This milling cycle can be repeated several times to maximize dispersion of particles. The narrow gaps (controllable from 500 to 5 μm) between the rollers, combined with the mismatch in angular velocity of the adjacent rollers, result in locally high shear forces with a short residence time. One of the unique advantages of calendering machines is that the gap width between the rollers can be mechanically or hydraulically adjusted and maintained, making it easy to get a controllable and narrow size distribution of particles in viscous materials. In some operations, the width of gaps can be decreased gradually to achieve the desired level of particle dispersion[19].

Calendering is widely used to mix electronic thick film inks, high-performance ceramics, carbon/graphite, paints, printing inks, pharmaceuticals, chemicals, glass coatings, dental composites, pigment, coatings, adhesives, sealants, and foods. The employment of a calendering technique to disperse CNTs in polymer matrices has become a very promising approach to achieve relatively uniform dispersion without causing severe damage to CNTs according to some recent reports[20,21]. The high shear stresses created between the narrow roller gaps can disentangle CNT bundles and distribute the individualized CNTs into the polymer matrix, while the short residence time likely limits the breakage of CNTs[21].

However, there are also several concerns about using this technique. One concern is that the gap width between the rollers is much larger than the size of individual CNTs, suggesting that this technique can only disentangle the CNT agglomerates into small ones on the microscopic scale, although some individual CNTs can be separated during the process (Figure 2.7). In

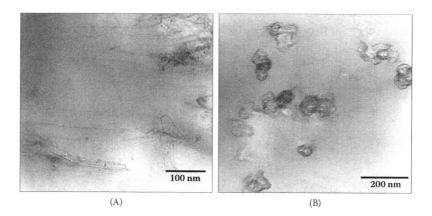

(A)   (B)

FIGURE 2.7
TEM images of nanoparticles dispersed by calendering machine in an epoxy matrix: (A) amino functionalized CNTs, and (B) carbon black. Note that both well-dispersed and agglomerated nanoparticles coexist in the nanocomposites. Adapted from Thostenson, E. T., et al. 2006. *Carbon* 44: 3022.

addition, the feeding material should be in a viscous state after mixing with CNTs, thus limiting the employment of this tool to disperse CNTs into the majority of thermosets and some thermoplastic matrices, such as polyethylene, polypropylene, and polystyrene. For thermosetting matrices, CNTs can be dispersed into a liquid monomer or oligomer, and nanocomposites can be obtained via in-situ polymerization.

2.3.3 Ball Milling

Ball milling is a type of grinding method used to grind materials into extremely fine powder for use in paints, pyrotechnics, and ceramics. During milling, high pressure is generated locally due to the collision between the tiny, rigid balls in a concealed container, which is schematically shown by Figure 2.8. An internal cascading effect of balls reduces the material to finer powder. Different materials, including ceramic, flint pebbles, and stainless steel, are used as balls. Industrial ball mills can operate continuously by feeding at one end and discharging at the other end. High-quality ball mills can grind mixture particles to as small as 100 nm, enormously increasing the surface area of particles. Ball milling has been successfully applied to transform CNTs into nanoparticles[22], to generate highly curved or closed-shell carbon nanostructures from graphite[23], to enhance saturation lithium composition in SWCNTs[24], to modify the morphologies of cup-stacked CNTs[25], and to generate different carbon nanoparticles from graphitic carbon for hydrogen storage applications[26].

Ball milling of CNTs in the presence of chemicals can enhance their dispersability, and also introduce functional groups onto the CNT surface. Ma et al. reported a simple chemomechanical method for amino functionalization of CNTs using ball milling[27,28]. The results showed that the addition of ammonium bicarbonate (NH_4HCO_3) for ball milling of CNTs effectively promoted the disentangling process (Figure 2.9), and the

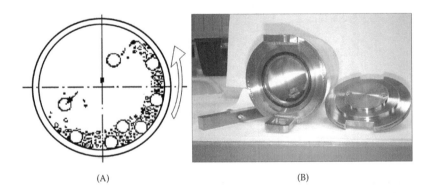

(A)   (B)

FIGURE 2.8
(A) Schematics of ball milling technique and (B) container.

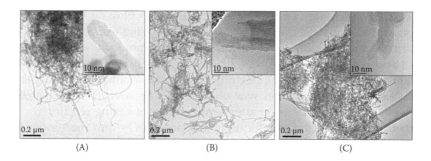

(A)   (B)   (C)

FIGURE 2.9
TEM images of CNTs milled after 9 hours: (A) as-received CNTs, (B) milled with NH_4HCO_3 and (C) milled without NH_4HCO_3. Adapted from Ma, P. C., et al. 2008. *Chem. Phys. Lett.* 458:166.

CNT length could be controlled by choosing an appropriate milling time. Useful functional groups, like amine and amide terminals, were attached to the CNT surface after milling[27]. The amino-functionalized CNTs can form covalent–bond CNTs with polymers and biological systems, as well as find potential applications in the electronics industries.

2.3.4 Stirring and Extrusion

Stirring is a common technique to disperse CNT particles in a solvent or polymer resin using a stirrer or shear mixer (Figure 2.10A). The size and shape of the propeller and the mixing speed control the dispersion quality[29]. A high-speed shear mixer with speed up to 10,000 rpm can be used to achieve fine dispersion of heavily agglomerated CNTs in a polymer matrix. MWCNTs

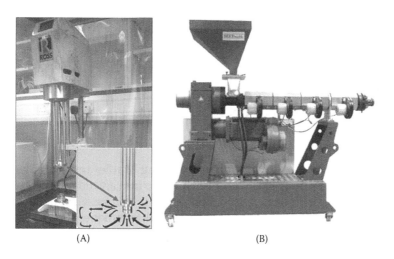

(A)   (B)

FIGURE 2.10
(A) Typical shear mixer and (B) extruder used for CNT dispersion.

can be dispersed more easily than SWCNTs by employing this technique, although the former CNTs tend to reagglomerate due to frictional contacts and elastic interlocking mechanisms[30]. Other parameters, such as sliding forces and weak attractive forces, have only marginal effects on final dispersion quality. However, agglomeration is almost spontaneous under static conduction. Reagglomeration was observed in apparently well-dispersed CNTs in an epoxy matrix even after several hours of curing reaction[6].

Extrusion is a popular technique to disperse CNTs into thermoplastics, especially for nanocomposites containing a high CNT content (Figure 2.10B). Thermoplastic pellets mixed with CNTs are fed into the extruder hopper and a twin screw rotates at a high speed to create a high shear flow, allowing dispersion of CNTs from agglomerates and mixing them into the polymer. Villmow et al. used this technique to disperse MWCNTs into poly(lactic acid), showing that the use of a distributive screw combined with a high rotation speed led to a high degree of CNT dispersion[31].

2.3.5 Other Techniques

The techniques developed for CNT dispersion are not limited to those described previously. Use of a combination of the aforementioned techniques, such as ultrasonication and ball milling[28] and ultrasonication and extrusion[32,33], has become popular. Indeed, there is no almighty tool to achieve a perfect dispersion of CNTs in a polymer matrix. Several important factors, such as the physical state of polymer matrix (solid or liquid), the CNT content, use of solvents, availability of techniques, and the fabrication process of composites, should be considered in selecting a proper technique. Table 2.2 compares the characteristic features of various dispersion techniques, thus providing a general guideline for the selection of proper techniques for specific end purposes.

2.4 Characterization of CNT Dispersion

2.4.1 Microscopic Methods

Besides the obvious difficulty in obtaining stable and homogeneous dispersions of CNTs, another complication is to find a valid method to evaluate their dispersion states. Dispersion of CNTs can be visualized directly by employing some microscopic and physical techniques, or indirectly by measuring the mechanical/electrical response of the composites containing CNTs. The latter will be discussed specifically in Chapter 4.

There are a few approaches that can be employed to evaluate the dispersion of CNTs either in a solvent or polymer matrix. The long-term stability of CNT

TABLE 2.2 Comparison of Various Techniques for CNT Dispersion in Nanocomposites.

Technique	Damage to CNTs	Suitable Polymer Matrix	Governing Factors	Availability
Ultrasonication	Yes	Soluble polymer, low viscous polymer or oligomer, monomer	Power and mode of sonicator, sonication duration	Commonly used in laboratories; easy operation and cleaning after use
Calendering	No. CNTs may be aligned in matrix	Liquid polymer or oligomer, monomer	Roller speed, distance between adjacent rolls	Operation training is necessary; hard to clean after use
Ball milling	Yes	Powder (polymer or monomer)	Milling time, rotation speed, size of balls, balls/CNT ratio	Easy operation; need to clean after use
High-speed shear mixing	Yes	Soluble polymer, low viscous polymer or oligomer, monomer	Size and shape of propeller, mixing speed and time, viscosity of liquid	Commonly used in lab; easy operation and cleaning after use
Extrusion	No	Thermoplastics	Temperature, configuration, and rotational speed of screw	Large-scale production; operation training is necessary; hard to clean after use

suspension is often used as a good indicator to show the dispersion of CNTs in various organic solvents and/or surfactants[34]. However, use of CNT suspensions cannot distinguish individual CNTs from CNT agglomerates that are dispersed in a solvent. Transmission-mode optical microscopy (OM) has been popularly used to visualize the degree of CNT dispersion in a polymer matrix. Figure 2.11 presents the morphologies of nanocomposites containing CNTs or carbon black (CB), illustrating different views depending on the magnification used[35]. The nanocomposites containing 0.2% CNTs showed a sufficiently high degree of dispersion at a low magnification (Figure 2.11A), whereas there were typical resin-rich regions and CNT agglomerates as seen from the transparent and dark parts of the images at a high magnification (Figure 2.11B). The distribution of CB particles in the matrix appeared to be more uniform than those containing CNTs on the macroscopic scale (Figure 2.11C). However, severe agglomerates were observed at a higher magnification (Figure 2.11D) although their sizes were relatively smaller than the CNT agglomerates. It is important to visualize the morphologies on different

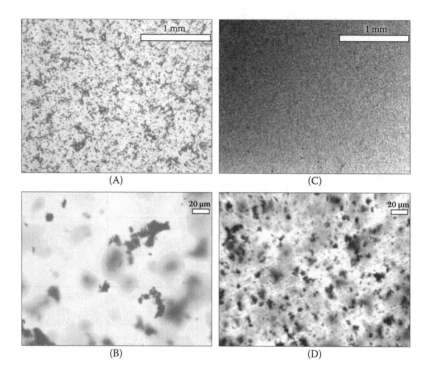

FIGURE 2.11
Transmission-mode optical microscope images of nanocomposites filled with different fillers at different magnifications: (A) and (B) 0.2% CNTs; (C) and (D) 0.2% CB. Adapted from Ma, P. C., et al. 2009. ACS *Appl. Mater. Interfaces* 1: 1090.

scales when employing OM to describe the dispersion states of CNTs in a polymer or solvent.

The atomic force microscope (AFM) is a very high-resolution microscope with a demonstrated resolution in the nanometer scale, which is more than 1000 times better than the limitation of OM. The information is gathered by "feeling" the surface of a substrate, usually a silicon wafer, on which CNTs are deposited. A very common option for AFM is to measure the bundle heights of CNTs, and a smaller average bundle size is an indication of better CNT dispersion[36]. Unfortunately, the bundle size can change as nanotubes are transferred onto the substrate, meaning that such a measurement does not necessarily reflect the state of dispersion in the original solvent.

The scanning electron microscope (SEM) and transmission electron microscope (TEM) are widely employed to characterize the structure and morphology of CNTs, as well as to evaluate the dispersion of CNTs either in solid state or in a polymer matrix (see Section 1.3 and Figure 1.11). It should be noted that imaging CNT dispersion by SEM often requires gold coating or carbon sputtering, which may adversely affect the accuracy of the morphologies. CNT solutions can be flash-frozen as a thin layer and are best viewed with cryogenic TEM, which is ideally suited for imaging of wet samples[37–39]. It is known that while the cryogenic TEM can maintain the dispersion states of CNTs in the original solution, its low resolution and stability may prevent one from truly observing the individual nanotubes.

2.4.2 Light Methods

The dispersion state of CNTs in liquids can also be evaluated based on several methods using light. One can observe the fraction of CNTs remaining after dispersion with naked eyes or via a digital photograph, although quantitative evaluation is rather difficult due to the black color of CNTs even at a very low concentration and lack of rating dispersion quality.

Characterization of CNT suspensions based on a particle size analyzer that uses a dynamic light-scattering technique is known to be the most popular for quantitative evaluation among various techniques. In this measurement, particle sizes are measured after dispersion in a solvent as the "effective diameter of CNT agglomerates" and the corresponding mean value and a standard deviation are recorded. Figure 2.12A shows a histogram of CNT dispersion measured using a particle size analyzer, reporting an average CNT size of 136 ± 92.4 μm. Li et al.[6] studied the effect of functionalization on the dispersion behavior of CNT using this technique, and found that the size distributions of CNT agglomerates produced by different functionalization techniques were in good agreement with the conclusions obtained from TEM observation. It is noted that the dynamic light-scattering technique can

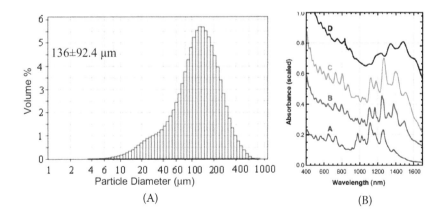

(A)   (B)

FIGURE 2.12
(A) Size distributions of CNT agglomerates obtained using a particle size analyzer, adapted from Li, J., et al. 2007. *Adv. Funct. Mater.* 17: 3207 and (B) SWCNT adsorption spectra in aqueous suspension of surfactant (sodium dodecyl sulfate), adapted from O'Connell, M. J., et al. 2002. *Science* 297: 593.

also be applied to spherical and disk-like particles, such as CB and graphite nanoplatelets[40,41].

The discovery of nanotube fluorescence has led to development of a precise method to detect dispersion of individual CNTs[42–45]. A nanotube in an aligned bundle does not emit fluorescence because of energy transfer to neighboring tubes, particularly to the metallic ones. Thus, the dispersion process can be monitored by examining transient fluorescent emission as a function of various parameters, like the type and concentration of the surfactant used, sonication time, and CNT functionalization. The spectral structures A, B, and C in Figure 2.12B are interpreted as a superposition of distinct electronic transitions from a variety of isolated nanotubes with different diameters. In contrast, broadened and red-shifted adsorption spectra D are typical of aggregated nanotubes.

2.4.3 Zeta Potential

Besides microscopic and light methods, the dispersion of CNTs can also be monitored based on a physical principle, such as zeta potential. From a theoretical viewpoint, zeta potential is an electric potential in the interfacial double layer at the location of the slipping plane versus a point in the bulk fluid away from the interface in a colloidal system. In other words, zeta potential measures the potential difference between the dispersion medium and the stationary layer of fluid attached to the dispersed particle. It indicates the degree of repulsion between adjacent, similarly charged particles in a colloid[46]. When the particles are sufficiently small, a high zeta potential corresponds to stable dispersion, that is, the dispersed particles resist aggregation. When the potential is low, attraction exceeds repulsion and the dispersion breaks and flocculates.

This method has been applied for the evaluation of CNT dispersion[47–49]. Recently, Ma et al. proposed a correlation among zeta potential, surface energy, and oxygen contents of functionalized CNTs to different degrees[50]. Zeta potentials of several different colloidal systems consisting of CNTs dispersed in different solvents, such as water, ethanol, and hexane, were measured to quantitatively evaluate CNT dispersion in these liquids. The results showed that the absolute zeta potential values increased consistently with an increasing degree of functionalization as indicated by the increase in the oxygen-to-carbon (O/C) ratio, and the increase in potential value was most pronounced in a hydrophilic liquid such as water (Figure 2.13). It was revealed that the CNT dispersability in a liquid was affected by the hydrophilicity of the CNT surface, which was reflected by the potential value because the surface functionality promoted by functionalization could resist aggregation. Judging from these observations, 25 mV was regarded as the threshold zeta potential, above which the CNT dispersion was stable. The reliability of this measurement was further confirmed by optical and electron microscopy.

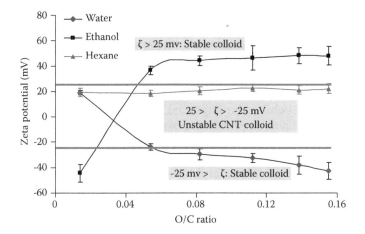

FIGURE 2.13
Correlation between Zeta potential, ζ, and CNT functionality in different solvents. Adapted from Ma, P. C., et al. 2010. *Carbon* 48: 1824.

2.5 Dispersion of CNTs in Liquid Media

2.5.1 Dispersion of CNTs in Water and Organic Solvents

The dispersion of CNTs either in water or in organic solvents may be necessary for their chemical and physical examination, since it allows easy characterization using different techniques. In addition, the dispersion of CNTs is important to fabricate CNT/polymer nanocomposites, as techniques based on predispersion of CNTs in solvent and following mixing with a monomer or polymer have been proved to be an effective way to enhance the properties of composites, especially for the fabrication of thermoset-based CNT/polymer composites. Therefore, intensive research has been addressed to study the dispersion behavior of CNTs either in water or in organic solvents.

The term *solubility* is often employed to describe the dispersion of CNTs in water or organic solvents, although from the point of strict definition, CNTs cannot be dissolved in the aforementioned media. Solubility of CNTs in various solvents, such as dimethylformamide (DMF), chloroform, acetone, toluene, benzene, and ethanol, was studied in terms of Hansen solubility parameters[51–54], which are defined as:

$$\delta_t = \delta_d + \delta_p + \delta_h \tag{2.1}$$

where δ_t is the total solubility parameter (unit $\text{MPa}^{0.5}$), and the subscripts d, p, and h refer to the dispersive, polar contributions and the contribution arising from the hydrogen-bond interactions between the CNTs and solvents, respectively. It was shown[52] that CNTs were suspended very well in solvents with a certain range of dispersive component value, $\delta_d = 17–18$ MPa$^{0.5}$, whereas they tended to precipitate in solvents with high polar and hydrogen-bonding components (δ_h, $\delta_p > 10$ MPa$^{0.5}$), as summarized in Table 2.3[51,52]. However, the dispersion state had no specific dependency on the total solubility parameter, δ_t.

The mechanism behind the improved CNT dispersion in some solvents with a high dispersive component is possibly due to the enhanced wettability between CNTs and solvents. A non-hydrogen-bonding Lewis base with a high dispersive component and a low hydrogen-bonding parameter, such as DMF, N-methylpyrrolidone, and hexamethylphosphoramide, have demonstrated the ability to wet the CNT surface, resulting in a stable CNT solution[55–57]. The wetting characteristics of CNTs could be significantly improved via chemical functionalization[53]. However, the high dispersive component alone was insufficient to keep a stable CNT dispersion, because solvents like benzene, dimethyl sulfoxide, and styrene are not effective mediums for CNT

TABLE 2.3 Dispersion of CNTs in Various Solvents with Different Values of Hansen Parameters of the Solvents

Parameter Organic Solvent	δ_d (MPa$^{0.5}$)	δ_p (MPa$^{0.5}$)	δ_h (MPa$^{0.5}$)	δ_t (MPa$^{0.5}$)	Dispersion State
DMF	17.4	13.7	11.3	24.8	Dispersed
Chloroform	17.8	3.1	5.7	19.0	Dispersed
1-Methyl-2-pyrrolidone	18.0	12.3	7.2	22.9	Dispersed
2-Propyl alcohol	15.8	6.1	16.4	23.5	Swollen
1-Pentyl alcohol	16.0	4.5	13.9	21.7	Swollen
Tetrahydrofuran	16.8	5.7	8.0	19.4	Swollen
Toluene	18.0	1.4	2.0	18.2	Swollen
o-Methoxyphenol	18.0	8.2	13.3	23.8	Swollen
Dichloromethane	18.2	6.3	6.1	20.3	Swollen
Benzene	18.4	0.0	2.0	18.6	Swollen
Dimethyl sulfoxide	18.4	16.4	10.2	26.7	Swollen
Styrene	18.6	1.0	4.1	19.0	Swollen
Methyl methacrylate	13.7	9.8	6.1	17.9	Sedimented
Methanol	15.1	12.3	22.3	29.6	Sedimented
Hexane	15.3	0.0	0.0	15.3	Sedimented
Acetone	15.5	10.4	7.0	20.0	Sedimented
Water	15.6	16.0	42.3	47.8	Sedimented
Ethanol	15.8	8.8	19.4	26.5	Sedimented
Acrylonitrile	16.4	17.4	6.8	24.8	Sedimented

Source: Adapted from Barton, A. F. M. 1991. *CRC handbook of solubility parameters and other cohesion parameters*, 95. Boca Raton, FL: CRC Press and Ham, H. T., et al. 2005. *J. Colloid Interface Sci.* 286: 216.

solubility even though their δ_p values are higher than 18 MPa$^{0.5}$. A systematic study of the efficacy of a series of amide solvents to disperse as-produced and purified CNTs suggested that favorable interactions between CNTs and alkyl amide solvents was attributed to the highly polar π system and optimal geometries with appropriate bond lengths and bond angles of the solvent structures[58].

2.5.2 Dispersion of CNTs in Polymers

Physical association of polymers with CNTs has been shown to enhance the dispersion of CNT in both water[34] and organic mediums[59], as well as to enable separation of nanotubes from carbonaceous and catalyst impurities[60]. Two mechanisms are introduced to explain these phenomena: (1) one is polymer wrapping[61], which relies on specific interactions between the CNT surface and the polymer via physical adsorption or chemical bonding and (2) another is the kinetic mechanism[62] in that the long-range entropic repulsion among polymer-decorated CNTs acts as a barrier preventing the CNTs from approaching and agglomerating with one another. Further details of topics that are related to CNT functionalization and processing of CNT/polymer nanocomposites will be discussed in Chapter 3.

2.6 CNT Dispersion Using Surfactants

2.6.1 Role of Surfactants in CNT Dispersion

Surfactants are wetting agents that decrease the surface tension of a liquid, allowing easier spreading and lower interfacial tension between two liquids or a liquid–solid interface. They are usually organic compounds that are amphiphilic, meaning they contain both hydrophobic groups and hydrophilic groups[63]. The hydrophilic region, called the *head group*, contains a polar structure, whereas the hydrophobic region, or the *tail group*, usually consists of one or a few hydrocarbon chains. Due to the distinct structural features, surfactants can be soluble in both organic solvents and water.

Surfactants are classified into ionic and nonionic surfactants according to the presence of formally charged groups in their heads[63]. A nonionic surfactant has no charge groups in its head. The head of an ionic surfactant carries a net charge. If the charge is negative, the surfactant is more specifically called *anionic*; if the charge is positive, it is called *cationic*. If a surfactant contains a head with two oppositely charged groups, it is termed *zwitterionic*.

There are a larger number of reports available on CNT dispersion using ionic surfactants than nonionic surfactants. This is possibly due to the following reasons: (1) ionic surfactants are excellent foamers, and thus can effectively improve the wettability between solvents and CNTs, (2) ionic surfactants exhibit excellent solubility in water, which is one of the most commonly used solvents for CNT dispersion, sorting, and characterization, and (3) ionic surfactants are inexpensive and easier and safer to use than nonionic surfactants. Table 2.4 compares the advantages and disadvantages of different surfactants, providing a general guideline for selecting proper surfactants for CNT dispersion.

There are two important features that describe the properties of surfactants used in processing stable colloidal dispersions, which are (1) adsorption at the interface and (2) self-accumulation into super-molecular structures[53,63–66]. The adsorption of surfactants onto inorganic and organic surfaces usually depends on the chemical characteristics of particles, surfactant molecules, and solvents. The driving force for the adsorption of ionic surfactants on charged surfaces is the coulombic attraction. The mechanism by which nonionic surfactants adsorb onto a hydrophobic surface is the strong hydrophobic attraction between the solid surface and the surfactant's hydrophobic tail. Once the surfactant molecules are adsorbed onto the particle surface, the surfactant molecules are self-assembled into micelles, aggregates of surfactant molecules dispersed in a liquid colloid, above a critical micelle concentration (CMC). The adsorption of surfactant on the CNT surface effectively lowers the surface tension, preventing the formation of aggregates.

TABLE 2.4   Advantages and Disadvantages of Different Surfactants.

Surfactant Feature	Ionic Surfactant			Nonionic Surfactant
	Anionic	**Cationic**	**Zwitterionic**	
Advantages	Inexpensive; high purity; completely ionized in water; excellent wetting ability and good stability in alkaline media; less irritating to skin and eyes	Compatible with nonionic and zwitterionic surfactants; strong adsorption onto solid surface; formation of emulsion	Compatible with other surfactants; less irritating to skin and eyes; adsorbable onto charged surface without forming hydrophobic film	Compatible with other surfactants; soluble in water and organic solvents; excellent dispersing agent for carbon; resistant to cations and electrolyte at high concentrations
Disadvantages	Insoluble in electrode and organic solvents; unstable in acid media	High cost, incompatible with anionic surfactant; poor suspending power to carbon	Insoluble in most organic solvents	Highly viscous; poor foamer; no electrical effect; mixture of products

Furthermore, surfactant-treated CNTs readily overcome van der Waals attractions by electrostatic/steric repulsive forces[64,65].

New insights into CNT dispersion on a molecular level have been offered recently through computer simulations[66–68] that localized interactions among surfactant molecules and CNTs. In equilibrium, a dynamic balance is established where a surfactant molecule can exist in one of the three equal-energy states: as an isolated individual molecule, in micelles in the solution, or adsorbed onto the CNT surface. The balance is sensitive to the concentration of surfactant. In diluted solutions, the adsorption of surfactant onto CNTs appears to be random, while when the concentration is high, most surfactants form hemispheres on the CNTs, and at an intermediate concentration, both situations coexist. Another interesting observation is that tube junctions are favored for surfactant adsorption due to the apparently reduced energy penalty for more flexible tail conformation[68]. Different surfactants could behave differently because of varying interaction strengths and CMCs, while the micelle concentration depends on the structure of surfactant[69,70]. If surfactants with long saturated hydrocarbon tails are used, they would adsorb onto CNTs predominantly aligned along the tube axis, implying the onset of template crystallization as observed for long polymer chains on CNTs. On the other hand, a specific surfactant could have different interaction energies with CNTs of different diameters, a subtle effect that could be potentially useful for sorting CNTs according to their diameters.

A typical dispersion process involves adsorption of surfactant onto the CNT surface under ultrasonication for minutes or hours. The outermost nanotubes in a bundle tended to be treated more than the innermost CNTs and in some cases the CNTs remained predominantly agglomerated even after surfactant treatment[44,64], indicating the need for mechanical exfoliation of the bundles prior to surfactant treatment if individual CNTs are to be separated. A mechanism of nanotube isolation from a bundle with the combined assistance of ultrasonication and surfactant adsorption is schematically shown in Figure 2.14A[43]. The role of ultrasonication is to apply a high local shear force to the nanotube bundle end (Stage ii). Once spaces or gaps at the bundle ends are formed, they propagate along the nanotube length by surfactant adsorption (Stage iii), ultimately separating the individual CNTs from the bundle (Stage iv).

Dispersion of CNTs in an aqueous solution containing a surfactant works mainly through hydrophobic and hydrophilic interactions, in which the hydrophobic tail of the surfactant molecule adsorbs on the surface of individual or bundled CNTs while the hydrophilic head associates with water for dissolution[70]. Visualization of how the adsorbed amphiphilic molecules organize on the CNT surface is challenging, but there are three probable configurations that describe the interactions between individual CNTs and surfactants, as schematically shown in Figure 2.14B[69]. CNTs can be encapsulated within cylindrical micelles or covered with either hemispherical micelles or randomly adsorbed molecules.

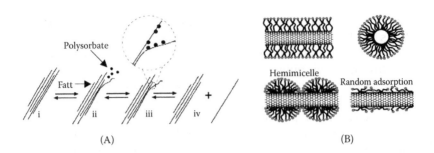

FIGURE 2.14

(A) Mechanism of individual nanotube isolation from a bundle via by ultrasonication and surfactant stabilization, adapted from Strano, M. S., et al. 2003. *J. Nanosci. Nanotechnol.* 3: 81 and (B) schematic illustrations of surfactant assembly on a CNT, adapted from Wallace, E. J., et al. 2009. *Nanotechnology* 20: 045101.

Comprehensive studies based on experiments and simulations have been carried out to measure the interactions between modified AFM tips and surfactant-treated CNTs[70,72]. Studies have also been made to qualitatively assess preferred interactions by testing dispersion with various surfactants[71]. The results showed that the surfactants with polar end groups were more attractive to SWCNTs than nonpolar ones, and the CNTs could behave as both donors and acceptors according to the chemical states of the medium. In the presence of strong donors such as benzene rings, CNTs behave as acceptors; and with strong acceptor groups, they act as donors. Thus, the attraction is maximal when both are present[70]. Ionic surfactants in general favor CNT/water soluble solutions. Alternatively, nonionic surfactants are preferred if organic solvents have to be used. Further discussion on surfactant-assisted dispersion of CNTs is provided in the following text.

2.6.2 Nonionic Surfactant-Assisted CNT Dispersion

A nonionic surfactant has no charge group in its head. The head groups can consist of functional groups like hydroxyl (–OH) or amine terminals. Figure 2.15 shows the chemical structures of commonly used nonionic surfactants for CNT dispersion. As discussed in Section 2.6.1, nonionic surfactants are adsorbed onto the hydrophobic CNT surface based on a strong hydrophobic attraction. The nature of the surfactant, its concentration, and type of interaction play a crucial role in both the phase behavior of CNT dispersion[63] and the properties of CNTs and CNT/polymer composites.

Only limited research work has thus far been reported on nonionic surfactant-assisted CNT dispersion in organic solvents as compared to water-soluble systems[67]. Unlike in aqueous solutions, hydrophobic CNTs are easily wetted by organic solvents, and thus they tend not to self-assemble in bundles. The first report on CNT dispersion using a nonionic surfactant dates back to

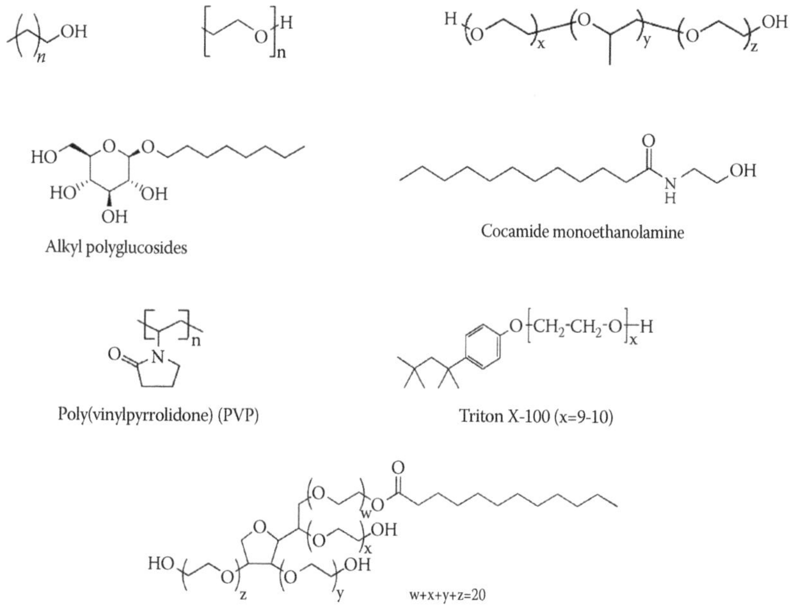

FIGURE 2.15
Chemical structures of commonly used nonionic surfactant.

2000[73] where a poly(ethylene oxide)-based surfactant (polyoxyethylene 8 lauryl) was dissolved in acetone and magnetically mixed with CNTs, which were then incorporated into an epoxy to fabricate composites. The addition of only 1 wt% surfactant-treated CNTs in the composite increased the glass transition temperature from 63 to 88°C along with a more than 30% increase in elastic modulus. In contrast, the addition of CNTs without surfactant had moderate effects on these properties. However, the dispersion states of CNTs in organic solvents were not investigated in this study.

Geng et al. systematically studied the effects of the concentration of nonionic surfactant on CNT dispersion and the corresponding mechanical properties of CNT/epoxy composites[74]. Triton X-100 with a CMC of 0.2 mM at room temperature was employed as a surfactant and two different concentrations of 1 and 10 CMC were studied as the working concentration of Triton X-100 was known to be in the range of 1–5 mM previously[75]. Figure 2.16 shows the schematic of a Triton X-100 molecule and micelle, and the corresponding interactions with CNTs. The long tail represents the hydrophobic segment, while the other end corresponds to the hydrophilic head (Figure 2.16A). At a CMC, the interface becomes saturated (Figure 2.16B), and above the CMC, the molecules start to form aggregates or micelles (Figure 2.16C). Because the larger was the micelle size, and the stronger was the steric repulsive force introduced by micelles (Figure 2.16D), it is expected that the surfactant with a higher concentration can more effectively disentangle large CNT

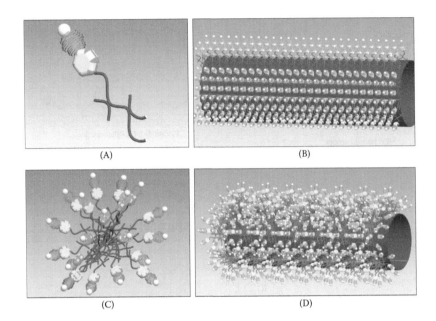

FIGURE 2.16
Schematics of (A) single Triton X-100 molecule, (B) a CNT wrapped by Triton X-100 molecules
(1 CMC), (C) a Triton X-100 micelle, and (D) a CNT wrapped by Triton X-100 micelles (10 CMC).
Adapted from Geng, Y., et al. 2008. *Compos. A* 39: 1876.

agglomerates. The attachment of Triton X-100 to a CNT surface was confirmed using characterization tools like Fourier transform infrared (FT-IR)
spectroscopy and x-ray photoelectron spectroscopy (XPS).

Comparing the stability of CNTs in a solvent is one of the most direct means
to understand the effects of surfactant treatment. The dispersion states of a
CNT/acetone suspension indicates excellent stability of the Triton-treated
CNTs for up to 48 hours. Pristine CNTs showed rapid precipitation after 5 minutes of sonication, along with reaggregation during the sedimentation process.
The surfactant molecules or micelles that were adsorbed onto the CNT surface
were not only able to decrease the surface tension, but also separate the large
agglomerates by overcoming the van der Waals attraction, so as to maintain
good stability. There was little difference between the treated CNTs with different micelle concentrations, indicating a low CMC requirement to achieve
stable dispersion of CNTs in acetone. TEM allowed further evaluation of the
differences in dispersion state introduced by surfactant treatment as shown
in Figure 2.17. Large agglomerates and closely packed CNTs were prevalent
in the pristine state (Figure 2.17A), which became significantly loosened after
the surface treatment without breakage or shortening of CNTs (B and C in
Figure 2.17). The use of a high concentration (10 CMC) gave rise to an even
better dispersion state than the low concentration (1 CMC).

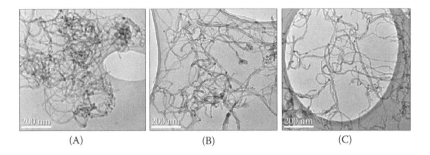

(A)   (B)   (C)

FIGURE 2.17
TEM images of (A) pristine, (B) Triton X-100 (1 CMC) treated, and (C) Triton X-100 (10 CMC) treated CNTs. Adapted from Geng, Y., et al. 2008. *Compos. A* 39: 1876.

The improved CNT dispersability in organic solvent and attachment of surfactant on the CNT surface resulted in both improved CNT dispersion in a polymer matrix and enhanced interfacial interaction between CNTs and the matrix. The corresponding mechanical, fracture, and thermomechanical properties all displayed significant improvements after the surfactant treatment of CNTs, indicating that the surfactant served as an effective dispersant and an interfacial coupling agent.

2.6.3 Ionic Surfactant-Assisted CNT Dispersion

Ionic surfactants refer to those containing a charged head. Figure 2.18 shows the chemical structures of commonly used ionic surfactants. They can be further classified into anionic, cationic, and zwitterionic surfactants. The surface-active portion of the molecule in an anionic surfactant bears a negative charge. Organic compounds containing sulfate, sulfonate, or carboxylate anions are typical examples of anionic surfactants (A–F in Figure 2.18). Cationic surfactants consist of the surface active portion of the molecules bearing a positive charge (G–J in Figure 2.18). Most commercially available cationic surfactants contain nitrogen and halogen atoms. When both negative and positive charges are present in the surface active portion of the molecules, they are called zwitterionic surfactants (K–L in Figure 2.18).

As for the nonionic surfactants, the effectiveness of ionic surfactants in the processing of CNT dispersion in aqueous media depends largely on their type and concentration. The first paper on CNT dispersion using an ionic surfactant was reported by Bandow et al. in 1997[76]. The cationic surfactant, benzalkonium chloride, with a concentration of 0.1% in aqueous solution, was employed to disperse and purify SWCNTs prepared by laser ablation. The results showed that a cationic surfactant was a good candidate to separate coexisting carbon nanospheres, metal nanoparticles, polyaromatic

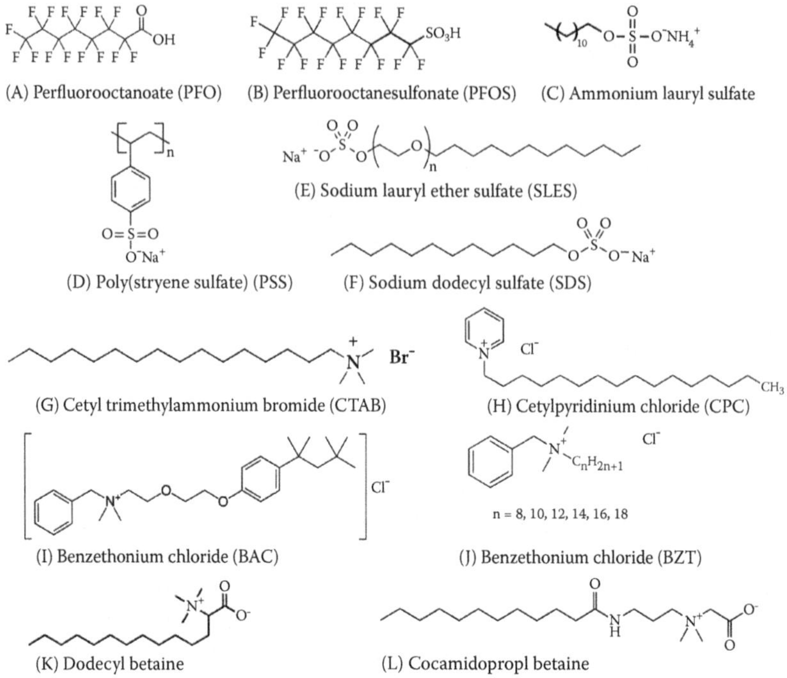

FIGURE 2.18
Chemical structures of commonly used ionic surfactants for CNT dispersion.

carbons, and fullerenes from the SWCNT fraction, and the purity of SWCNTs at the final stage sample was in excess of 90% by weight.

Among many ionic surfactants, sodium dodecyl sulfate (SDS) and sodium dodecylbenzene sulfonate (SDBS) were commonly used to improve CNT dispersion and decrease the tendency of dispersed CNTs to form reagglomeration in water[53]. Islam et al. evaluated various SWCNT aqueous suspensions containing the above surfactants with optimum nanotube-surfactant ratios of 1:5 to 1:10 by weight[77]. The results showed that the SWCNTs suspended with SDS at a concentration of 0.1 mg/ml was stable for less than a week, whereas the SWCNT solution containing up to 20 mg/mL of SDBS was stable for the same period. The AFM evaluations further indicated that the SDBS-assisted SWCNT suspension contained many individual SWCNTs even at a relatively high concentration of 10 mg/mL, and was stable over months without significant aggregation or bundling. The strong interactions between the SDBS surfactant and SWCNTs arose from the combined effects of the long lipid chains of SDBS and the π-π stacking between aromatic rings on the surfactant molecule and the graphitic surface of nanotubes. Such hydrophobic and π-π interactions have also been confirmed in another study[78] where SWCNTs were dispersed in an aqueous solution containing a cationic surfactant (trimethyl-(2-oxo-2-pyren-1-yl-ethyl)-ammonium bromide) under mild soni-cation in water. It was shown that both aromatic and naphthenic (saturated rings) groups offered good affinity between the surfactant and CNTs[53,79].

The SDS molecules, as well as many other synthetic amphiphilic molecules with long lipid chains, are also known to effectively adsorb onto CNTs having cylinder-like structures[80]. As such, the use of SDS surfactants has stimulated significant research interest, especially for dispersion, chemical manipulation of SWCNTs, as well as sorting CNTs with specific chiralities and electronic properties[3,81]. In addition to sulfate-based surfactants, other types of surfac-tants were also able to suspend individual SWCNTs in mass conversions on the order of around 5%[37]. Results from molecular simulation[42] suggested that the density of the surfactant-assisted SWCNT micelle is very close to that of water, which may explain the kinetically stable nature of CNT suspension in an aqueous medium.

Shin et al.[82] carried out the first example of a systematic study on the dispersion of SWCNTs in an aqueous solution by comparing the surfactant type and considering the effect of significant parameters such as the surfac-tant and SWCNT concentrations. Three different surfactants, that is, Igepal CO-990 (polyoxyethylene (100) nonylphenyl ether, nonionic), cetyltrimeth-ylammonium bromide (CTAB, cationic), and SDS (anionic), are employed for dispersing a high concentration of individual SWCNTs in an aqueous solu-tion. The results showed that all three surfactants could disperse SWCNTs well at the optimum concentration, which was found to be slightly higher than that of their CMC values. At very high concentration, SWCNTs were aggregated and formed bundles in all the cases. Therefore, by adding more surfactant than the optimum concentration means wasting material and may increase the cost of the process. It was also found that at a very minimum concentration, the ability of the Igepal to suspend SWCNTs was much better than that of cetyltrimethylammonium bromide (CTAB), which in turn was better than the SDS.

References

1 Thostenson, E. T., et al. 2001. *Compos. Sci. Technol.* 61: 1899.
2 Ajayan, P. M., et al. 2003. *Nanocomposite science and technology*, 77. New York: Wiley-VCH.
3 Lin, Y., et al. 2004. *J. Mater. Chem.* 14: 527.
4 Coleman, J. N., et al. 2006. *Adv. Mater.* 18: 689.
5 Fiedler, B., et al. 2006. *Compos. Sci. Technol.* 66: 3115.
6 Li, J., et al. 2007. *Adv. Funct. Mater.* 17: 3207.
7 Ma, P. C. 2008. Novel surface treatment, functionalization and hybridization of carbon nanotubes and their polymer-based composites. PhD dissertation. Hong Kong University of Science and Technology 8–10.

8 Kim, J. K., et al. 1998. *Engineered interfaces in fiber reinforced composites,* 1–100. New York: Elsevier.
9 Thess, A., et al. 1996. *Science* 273: 483.
10 Breuer, O., et al. 2004. *Polymer Compos.* 25: 630.
11 Xie, X. L., et al. 2005. *Mater. Sci. Eng. R.* 49: 89.
12 Moniruzzaman, M., et al. 2006. *Macromolecules* 39: 5194.
13 Lau, K. T., et al. 2006. *Compos. B* 37: 425.
14 Gibson, R. F., et al. 2007. *Compos. Sci. Technol.* 67: 1.
15 Bal, S., et al. 2007. *Bull. Mater. Sci.* 30: 379.
16 Sonifer Products. http://www.sonifier.com, Branson Ultrasonics Corp., (accessed June 2010).
17 Lu, K. L., et al. 1996. *Carbon* 34: 814.
18 Mukhopadhyay, K., et al. 2002. *Carbon* 40: 1373.
19 EXAKT. Three roll mills, http://www.exakt.com/Three-Roll-mills.25+M52087573 ab0.0.html (accessed June 2010).
20 Gojny, F. H., et al. 2004. *Compos. Sci. Technol.* 64: 2363.
21 Thostenson, E. T., et al. 2006. *Carbon* 44: 3022.
22 Li, Y. B., et al. 1999. *Carbon* 37: 493.
23 Gao, B., et al. 2000. *Chem. Phys. Lett.* 327: 69.
24 Huang, J. Y., et al. 1999. *Chem. Phys. Lett.* 303: 130.
25 Kim, Y. A., et al. 2002. *Chem. Phys. Lett.* 355: 279.
26 Awasthi, K., et al. 2002. *Intl. J. Hydrogen Energy* 27: 425.
27 Ma, P. C., et al. 2008. *Chem. Phys. Lett.* 458: 166.
28 Ma, P. C., et al. 2009. *J. Nanosci. Nanotechnol.* 9: 749.
29 Sandler, J. K. W., et al. 1999. *Polymer* 40: 5967.
30 Schmid, C. F., et al. 2000. *Phys. Rev. Lett.* 84: 290.
31 Villmow, T., et al. 2008. *Polymer* 49: 3500.
32 Moniruzzaman, M., et al. 2006. *Polymer* 47: 293.
33 Isayev, A. I., et al. 2009. *Polymer* 50: 250.
34 Bandyopadhyaya, R., et al. 2002. *Nano. Lett.* 2: 25.
35 Ma, P. C., et al. 2009. ACS *Appl. Mater. Interfaces* 1: 1090.
36 Hamon, M. A., et al. 2001. *Appl. Phys. A* 74: 333.
37 Moore, V. C., et al. 2003. *Nano. Lett.* 3: 1379.
38 Kedem, S., et al. 2005. *Langmuir* 21: 5600.
39 Dror, Y., et al. 2005. *Macromolecules* 38: 7828.
40 Ivakhnenko, V., et al. 2006. *J. Quant. Spectrosc. Radiat. Transfer.* 100: 165.
41 Pecora, R. 1984. *Pure Appl. Chem.* 56: 1391.
42 O'Connell, M. J., et al. 2002. *Science* 297: 593.
43 Strano, M. S., et al. 2003. *J. Nanosci. Nanotechnol.* 3: 81.
44 Dyke, C. A., et al. 2003. *Nano. Lett.* 3: 1215.
45 Ju, S. Y., et al. 2008. *Nature Nanotechnol.* 3: 356.
46 American Society for Testing and Materials, ASTM Standard D 4187-82.
47 Vaisman, L., et al. 2006. *Adv. Funct. Mater.* 16: 357.
48 Lee, J., et al. 2007. *Meas. Sci. Technol.* 18: 3707.
49 Sun, Z., et al. 2008. *J. Phys. Chem. C* 112: 10692.
50 Ma, P. C., et al. 2010. *Carbon* 48: 1824.
51 Barton, A. F. M. 1991. *CRC handbook of solubility parameters and other cohesion parameters,* 95. Boca Raton, FL: CRC Press.
52 Ham, H. T., et al. 2005. *J. Colloid Interface Sci.* 286: 216.

53 Vaisman, L., et al. 2006. *Adv. Colloid. Interface Sci.* 128–130: 37.
54 Chen, Q. H., et al. 2008. *J. Phys. Chem.* 112: 20154.
55 Ausman, K. D., et al. 2000. *J. Phys. Chem.* 104: 8911.
56 Furtado, C. A., et al. 2004. *J. Am. Chem. Soc.* 126: 6095.
57 Bahr, J. L., et al. 2001. *Chem. Commun.* 2: 193.
58 Landi, B. J., et al. 2004. *J. Phys. Chem. B* 108: 17089.
59 Kang, Y., et al. 2003. *J. Am. Chem. Soc.* 125: 5650.
60 Yudasaka, M., et al. 2000. *Appl. Phys. A* 71: 449.
61 O'Connell, M. J., et al. 2001. *Chem. Phys. Lett.* 342: 265.
62 Shvartzman-Cohen, R., et al. 2004. *J. Am. Chem. Soc.* 126: 14850.
63 Rosen, M. J. 2004. *Surfactants and interfacial phenomena*, 1–10. New York: John Wiley & Sons.
64 Shvartzman-Cohen, R., et al. 2004. *Langmuir* 20: 6085.
65 Grossiord, N., et al. 2006. *Chem. Mater.* 18: 1089.
66 Tummala, N. R., et al. 2009. *Acs. Nano.* 3: 595.
67 Chiu, C. C., et al. 2009. *Biopolymers* 92: 156.
68 Angelikopoulos, P., et al. 2008. *J. Phys. Chem. B* 112: 13793.
69 Wallace, E. J., et al. 2009. *Nanotechnology* 20: 045101.
70 Wang, H. 2009. *Current Opinion Colloid Interface Sci.* 14: 364.
71 Xie, H., et al. 2008. *J. Pept. Sci.* 14: 139.
72 Poggi, M. A., et al. 2004. *Nano. Lett.* 4: 61.
73 Gong, X., et al. 2000. *Chem. Mater.* 12: 1049.
74 Geng, Y., et al. 2008. *Compos. A* 39: 1876.
75 Barbero, M. C., et al. 1984. *Arch. Biochem. Biophys.* 228: 560.
76 Bandow, S., et al. 1997. *J. Phys. Chem. B* 101: 8839.
77 Islam, M. F., et al. 2003. *Nano. Lett.*; 3: 269.
78 Nakashima, N., et al. 2002. *Chem. Lett.* 31: 638.
79 Tan, Y., et al. 2005. *J. Phys. Chem. B* 109: 14454.
80 Richard, C., et al. 2003. *Science* 300: 775.
81 Hersam, M. C. 2008. *Nature Nanotechnol.* 3: 387.
82 Shin, J. Y., et al. 2008. *Chem. Eur. J.* 14: 6044.

3

Functionalization of CNTs

3.1 Introduction

Carbon nanotubes (CNTs) in their as-produced forms are difficult to disperse and dissolve in water or organic media, and they exhibit extremely high resistance to wetting. Difficulties also arise from incorporating the insoluble nanotubes into polymers when fabricating composites. This is because the carbon atoms on CNT walls are chemically stable due to the aromatic nature of the bonding. As a result, the reinforcing CNTs interact with the surrounding polymer matrix, mainly through van der Waals interactions, which may not provide an efficient load transfer across the CNT/matrix interface. It is known that the performance of a CNT/polymer nanocomposite depends critically on the dispersion of CNTs and the interfacial interactions between the CNTs and polymer. These challenges demand development of effective methods to modify the surface properties of CNTs. A suitable functionalization via the physical or chemical attachment of functionalities onto the CNT surface represents a strategy not only for improving CNT dispersion, solubility, and processability, but also for allowing strong interfacial interactions to take place between the CNTs and the polymer matrix. The endowed functionalities on CNTs can also be used to tailor the interactions of CNTs with other entities besides polymers, such as a solvent, biomolecules, and other nanoparticles. In addition, functionalized CNTs may exhibit mechanical or electrical properties that are different from those of pristine CNTs, and thus may be utilized for fine-tuning the chemistry and physics of CNTs[1]. Therefore, the functionalization of CNTs has been an attractive target for synthetic chemists and materials scientists for a long time.

Proper functionalization of CNTs is not easy due to the inherently inert nature of carbon atoms. CNTs possess two distinct regions, end tips and sidewalls, with different chemical reactivities. The presence of pentagons at the end tips of CNTs is dynamically more reactive than the hexagons on the sidewalls[2]. Therefore, functionalization of the sidewall comprising a regular graphene framework is more difficult to accomplish. The addition reactions to the partial carbon–carbon double bonds of hexagons cause the transformation of sp^2- into sp^3-hybridized carbon atoms, which is associated with a change from a trigonal-planar local bonding geometry to a tetrahedral geometry, and this process is energetically favorable at the end tips of CNTs[3].

The curvature of CNTs is an important factor determining their chemical or physical reactivity. The nonzero curvature of CNT end tips makes the tips more reactive than a planar graphene sheet. By the same token, the binding energy of atoms or functional groups on the sidewall of CNTs tends to increase with decreasing tube diameter because of the change in curvature. This tendency is supported by theoretical studies of the bond between the alkyl radicals and the sidewall of a single-walled CNT (SWCNT)[4]. In contrast, the concave curvature of the inner surface of a nanotube imparts a very weak reactivity toward addition reactions[5], so that CNTs are considered as nanocontainers or templates for some guest molecules to produce nanohybrids with novel properties.

This chapter is dedicated to studying the functionalization of CNTs. When dealing with functionalization of CNTs, a distinction must be made between the covalent and noncovalent functionalization. The working principles of these methods are discussed along with their effects on dispersion characteristics, as well as the mechanical, electrical/electronic, and thermal properties of CNTs. Nanoparticle/CNT hybrid materials are particularly useful to integrate the properties of the two components for use in catalysis, energy storage, and other nanotechnologies. The latest development in this field is also reviewed.

3.2 Covalent Functionalization of CNTs

3.2.1 Direct Sidewall Functionalization

Covalent functionalization is based on covalent linkage of functional entities onto the nanotube's carbon scaffold. It can be achieved either at the termini of the tubes or at their sidewalls. So far, it has not been possible to make quantitative assessments of whether such functionalization takes place preferentially on intact regions of the sidewalls or at preexisting defect sites. Direct covalent sidewall functionalization involves a change of hybridization from sp^2 to sp^3 and a simultaneous loss of conjugation. The typical diameter of a CNT is in the range of 1–20 nm, which is much smaller than that of the flat size of graphite (usually in micron scale), resulting in a low chemical reactivity for the CNTs. There are some other problems for reactions with CNTs, such as their low solubility or dispersability and the occurrence of CNTs in bundles or agglomerates. Therefore, functionalization of CNT sidewalls through the formation of covalent bonds can be possible only when a highly reactive reagent is used. Figure 3.1 illustrates various direct functionalization techniques and the corresponding chemical structures. Their principles and the conditions for the functionalization reactions are discussed in the following text.

3.2.1.1 Fluorination

Fluorination has been considered as one of the first sidewall functionalization processes (see Reaction A in Figure 3.1)[6], followed by a series of papers on the effects of CNT fluorination[7–12]. Using a fluorine gas, CNT sidewalls can be nondestructively fluorinated into C_2F at temperatures between 150 and 400°C[8]. It was noted that at temperatures above 400°C the graphitic structure of CNTs tended to decompose appreciably[8]. The highest degree of functionalization was obtained when CNTs were converted into CF using iodine pentafluoride IF_5[7]. However, extreme care must be exercised when SWCNTs of a small diameter are to be fluorinated because they are easily cut into small pieces by the etching effect of fluoride[12].

The fluorination reaction is very useful because further substitution can be done: alkyl groups could replace the fluorine atoms using Grignard[13] or organolithium[14] reagents. The alkylated CNT are well dispersed in common organic solvents such as tetrahydrofuran (THF) and can be completely dealkylated upon heating at 500°C in an inert atmosphere, thus recovering

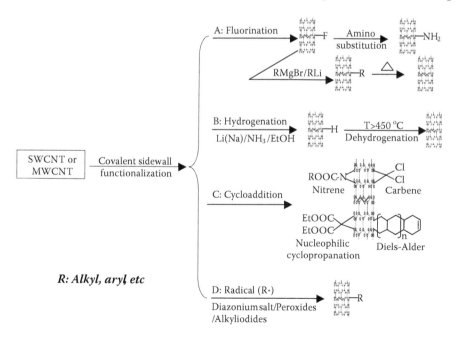

FIGURE 3.1
Direct sidewall functionalization of CNTs using different techniques.

pristine CNTs. In addition, several diamines were reported to react with fluorinated nanotubes via nucleophilic substitution reactions[15]. Because of the presence of terminal amino groups, these CNT are soluble in diluted acids and water. The amino-functionalized CNT were further modified, for example, by condensation with dicarboxylic acid chlorides, and can be employed to further bind various biomolecules to the sidewalls of CNTs for biological applications.

Halogenations of CNTs using other halogen elements, such as chlorine and bromine, have also been studied. For example, by electrolytic evolution of chlorine on an anode made from a foil of CNTs[16], the halogen atoms were coupled to the nanotube lattice. Oxygen-bearing functional groups, such as hydroxyl and carboxyl, were also formed simultaneously, which further enhanced the solubility of CNTs in water and alcohol. The functionalized CNTs can be converted with sodium amide or triphenylmethyl lithium to add other functional groups. The low susceptibility of CNTs to bromination was utilized as a means to purify multiwalled CNTs (MWCNTs) contaminated by amorphous carbon and carbon nanoparticles[17].

3.2.1.2 Hydrogenation

Several methods have been developed for hydrogenation of CNTs via dissolved metal reductions (see Reaction B in Figure 3.1). Chen et al. first reported the Birch reduction (organic reduction of aromatic rings with lithium and alcohol in liquid ammonia) of SWCNTs using lithium in diaminoethane[18]. The hydrogenated derivatives were thermally stable up to 450°C, while above this temperature a characteristic decomposition took place. Transmission electron microscopy (TEM) micrographs showed corrugation and disorder of CNT walls due to hydrogenation and the formation of CH bonds. The average hydrogen content determined from the thermogravimetric analysis-mass spectroscopy (TGA-MS) of hydrogenated SWCNTs using lithium in ammonia approximately corresponding to $C_{11}H_{19}$. Using sheets of SWCNT buckypaper as a negative electrode in an electrochemical cell containing an aqueous solution of KOH as electrolyte, Owens et al.[20] incorporated up to 6 wt% of hydrogen into the tubes. In theory, attaching a single hydrogen atom to any CNT is always exothermic, and zigzag nanotubes are more likely to be hydrogenated than armchair tubes with equal radius[21]. Indeed, hydrogenation also occurs even in the inner tubes of MWCNTs, as evidenced by the chemical composition and overall corrugation.

3.2.1.3 Cycloaddition

A cycloaddition is a pericyclic chemical reaction in which two or more unsaturated molecules, or part of the same molecule, combine with the formation of a cyclic adduct, allowing a net reduction of the bond multiplicity.

Cycloadditions are usually described by the backbone size of the participants. Cycloaddition to pristine CNTs (see Reaction C in Figure 3.1) was studied in an effort to functionalize CNT sidewalls[22]. According to the chemical structures of precursor for reactions, there are several forms of cycloadditions, including carbene, nitrene, ylide, Diels-Alder, and nucleophilic cyclopropanation[1,3,23–25].

A *carbene* is an organic molecule containing a carbon atom with six valence electrons and having the general formula RR'C: (R and R' are organic groups). Carbene can be generated in situ using a mixture consisting of chloroform and potassium hydroxide or a phenyl(bromodichloro methyl)mercury reagent. The addition of dichlorocarbene onto CNTs induced changes in the x-ray photoelectron spectroscopy (XPS) and far-infrared spectra, whereas chemical analysis showed the presence of chlorine in the sample[24]. However, the degree of CNT functionalization was low as the detected chlorine amount was only about 1.6 wt%. Because of the impure starting materials and a large amount of amorphous carbon on CNTs, it is difficult to ascertain the site of reaction between the CNTs and carbene. Nevertheless, the covalent modification using carbene showed strong effects on the dispersion and electronic band structures of CNTs.

A *nitrene* (R-N:) is the nitrogen analog of a carbene and is a reactive intermediate involving many chemical reactions. Thus, nitrene cycloaddition often leads to a considerable increase in the solubility of CNTs in organic solvents, such as tetrachloroethane, dimethyl sulfoxide (DMSO), dichlorobenzene. The first step of this reaction was the thermal decomposition of an organic azide, which gave rise to alkoxycarbonylnitrene via nitrogen elimination. This was followed by nitrene cycloaddition to the CNT sidewalls, generating an alkoxycarbonylaziridino-CNT. A variety of organic functional groups, such as alkyl chains, dendrimers, and crown ethers, were successfully attached to CNTs[23,24]. The modified CNTs containing chelating donor groups in the addends (originated from the nitrogen atoms of nitrene) allowed complexation of metal ions, such as Cu and Cd. The nitrene cycloaddition resulted in the formation of derivative CNTs that were soluble in dimethyl sulfoxide or 1, 2-dichlorobenzene.

An *ylide* is a neutral dipolar molecule with a positive and a negative charge on adjacent atoms.

This material was also developed for the cycloaddition of CNTs. Three main steps were involved: (1) the chemical modification of the starting material to form an ylide and following reaction with CNTs, (2) the separation of the soluble adducts and reprecipitation using a solvent or nonsolvent, and (3) the thermal removal of the functional groups followed by annealing at an elevated temperature. The functionalized CNTs were found to be free of amorphous carbon, whereas the catalyst content was less than 0.5%. The treatment of SWCNTs using azomethine ylide in dimethylformamide (DMF) resulted in the formation of substituted pyrrolidine moieties on the SWCNT surface, and the functionalized CNTs were soluble in most common organic solvents. The approach could be applied to all types of CNTs, including as-produced and oxidized SWCNTs, SWCNTs with various diameters, and those prepared by different methods, as well as MWCNTs.

The *Diels-Alder* cycloaddition was performed for sidewall functionalization of CNTs[23–25]. The reaction involves four π-electrons of a 1, 3-diene and two π-electrons of dienophile. The active reagent was o-quinodimethane (generated in situ from 4, 5-benzo-1, 2-oxathiin-2-oxide), and the reaction was assisted by microwave irradiation. The feasibility of the Diels-Alder cycloaddition onto the sidewalls of a SWCNT was assessed by means of a molecular modeling approach[24]. The Diels-Alder reaction of 1, 3-butadiene on the sidewall of an armchair (5, 5) nanotube was not quite feasible, and the cycloaddition of quinodimethane was found viable due to the aromatic stabilization at the corresponding transition states and products.

Fullerenes are known to react easily with bromomalonates via *nucleophilic cyclopropanation*. A similar reaction was performed using purified SWCNTs and diethyl bromomalonate as an addend[3,23,24]. In this reaction, diethylbromomalonate functioned as a formal precursor of carbene. The cycloaddition to dispersed CNTs in an organic undecene resulted in surface functionalization of CNTs. In a subsequent step, CNTs reacted with 2-(methylthio) ethanol to give a thiolated material. The functional groups on the CNT surface were visualized by a tagging technique using chemical binding of gold nanoparticles.

3.2.1.4 Radicals

Radicals are atoms, molecules, or ions with unpaired electrons on an open shell configuration. The unpaired electrons cause them to be highly chemically reactive, thus radicals can be employed for CNT functionalization. Classical molecular dynamics simulations have been used to model the attachment of carbon radicals onto CNTs[26], showing a high probability of reactions of radicals on the sidewalls of CNTs. There are several ways to generate radicals. In situ chemical generation of radicals using diazonium salt is one of the most commonly used methods, which is also proven to be an effective way for sidewall functionalization of CNTs (see Reaction D in Figure 3.1) and thus to allow uniform dispersion of nanotubes in both organic and aqueous solutions[27–30]. The same reaction can also be performed under solvent-free conditions, offering the possibility of an efficient scale-up with moderate volumes.

Thermal and photochemical routes have also been applied to sidewall functionalization of CNTs with radicals. Alkyl or aryl peroxides were decomposed thermally and the resulting radicals, namely phenyl or lauroyl, were added to the graphitic network[31,32]. Alternatively, CNTs were heated in the presence of peroxides and alkyl iodides or treated with various sulfoxides, resulting in the attachment of radicals on CNTs[33].

The reductive intercalation of lithium ions onto the nanotube surface has been studied in an ammonia atmosphere or in polar organic solvents[34–36]. The negatively charged CNTs were found to exchange electrons with long chain alkyl iodides, leading to the formation of transient alkyl radicals. The latter were bonded covalently to the CNT surface. It was also shown that H, N, NH, and NH$_2$ radicals could be added to CNTs, resulting in the attachment of amino groups on the CNT surface[37].

3.2.2 Defect Functionalization

Defect sites in CNTs can be the open ends and holes in the sidewall, and pentagon and heptagon irregularities in the hexagon graphene framework. Oxygenated sites must also be considered as defects. As discussed in Chapter 1, defects on CNTs play an important role in functionalization of CNTs and can take advantage of chemical transformation.

Up to now, all known methods for CNT synthesis generate impurities. The main by-products are amorphous carbon and catalyst nanoparticles. The techniques used for purification of the as-produced CNTs to remove the impurities, such as HNO$_3$, H$_2$SO$_4$, or a mixture of them, or with oxidants such as KMnO$_4$, ozone, or reactive plasma, create defects on both sidewalls and end tips of CNTs during the oxidative process. The defects and radicals on CNTs created by oxidants are stabilized by bonding with carboxylic (–COOH) or hydroxyl (–OH) groups, as shown in Figure 3.2. These functional groups have rich chemistry and the CNTs can be used as precursors for further chemical reactions[1,3,23–25,38,39], details of which are discussed in the following text.

1. The carboxylic groups on CNTs can be activated by conversion into acyl chloride groups with thionyl chloride. The acyl chlorides can be transformed to carboxamides by *amidation* (see Reaction A in Figure 3.2), allowing the decoration of CNTs with aliphatic amines, aryl amines, amino acid derivatives, peptides, or amino-group-substituted dendrimers. Similarly, amidation of CNTs can be achieved using di-cyclohexylcarbodiimide (DCC) and dimethylamino-pyridine (DMAP) as dehydrating agents and allowing the direct coupling of amines and carboxylic functions under mild and neutral conditions.

2. The *esterification* of oxidized nanotubes (see Reaction B in Figure 3.2) has become one of the most popular ways of producing soluble CNTs either in organic solvents or in water. It has been shown that the solid-state reaction between oxidized nanotubes and taurine (2-aminoethanesulfonic acid) afforded water-soluble material. Tasis et al. successfully solubilized short-length nanotubes by attaching glucosamine moieties, whereas galactose- and mannose-modified nanotubes were also prepared[24]. The grafting was obtained by producing the acyl chlorides

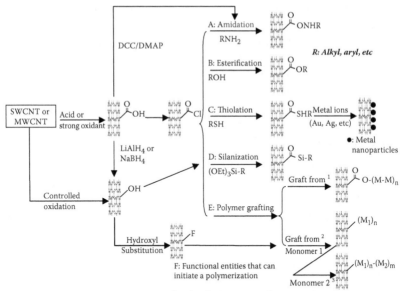

FIGURE 3.2
Defect functionalization of CNTs using different techniques.

or by carbodiimide activation, and the adducts were found to be water soluble. Carbohydrated CNTs were used to capture pathogenic *Escherichia coli* in solution for some bioapplications.

3. *Thiolation* of CNTs (see Reaction C in Figure 3.2) is particular useful as the thiol groups attached onto CNTs can effectively anchor biomolecules and metal particles, such as silver and gold nanoparticles, resulting in the production of new nanohybrids. The functionalization involves three major reactions, that is, carboxylation of CNTs using acids, chlorination using thionyl chloride, and thiolation using organic thiols (R-SH). Further details of producing metal/CNT nanohybrids using this process will be discussed in Section 3.7.

4. The hydrophobic surface of the MWCNTs was converted to a hydrophilic surface through UV/O_3 and *silanization* reaction (see Reaction D in Figure 3.2). This process is accomplished via the reaction of the –OH groups on CNTs (due to the UV/O_3 treatment) and the $-OCH_3$ groups of the silane molecules. The reaction leads to the improved CNT dispersion in organic solvent and attachment of silane molecules containing functional end groups on the MWCNT surface. This, in turn, contributed to enhancing the CNT's compatibility with polymer resins.

5. The defect functionalization of CNTs also leads to covalent *grafting of polymers* on CNTs. Basically, there are two kinds of processes to achieve this—*grafting to* and *grafting from* methods (see Reaction E in Figure 3.2). The *grafting to* method means that the readymade polymers with reactive end groups react with the functional groups on the nanotube surface. It is the reaction between the surface groups of nanotubes and the readymade polymers. Specifically, polymers terminated with amino or hydroxyl moieties are used in the amidation or esterification reactions with the carboxylic groups on the CNTs. Suitable polymers include, but are not limited to, poly(styrene-co-aminomethylstyrene), poly-(amic acid) containing a bithiazole ring, poly(propionylethylenimine-co-ethylenimine), monoamine-terminated poly(ethylene oxide) poly(styrene-cohydroxymethylstyrene), poly(styrene-co-p-(4-(40-inylphenyl)-3-oxabutanol)), poly(vinyl alcohol), poly(vinyl acetate-co-vinyl alcohol), poly[3-(2-hydroxyethyl)-2, 5-thienylene], and poly(ethylene glycol).

The *grafting from* method means the reactive groups are covalently attached to the CNT surface and then the polymers graft from the reactive groups. It is the reaction between the reactive groups on the surface of nanotubes and monomers. Therefore, the endowment of CNTs with specific functionalities that can initiate the polymerization becomes the key step for such a reaction. The grafting of polymers on CNTs using in situ atom transfer radical polymerization (ATRP) is a typical example of the grafting from approach. By immobilizing the ATRP initiators on CNTs, different polymers, such as polystyrene, poly(methyl methacrylate), and poly(*n*-butyl methacrylate) chains have been grown from the CNT surface[23,24].

It should be noted that the methods for defect functionalization of CNTs are not limited to the processes described above. Indeed, this topic is one of the most emerging areas in materials science and engineering. The nature of CNT surface functionalized in either of the previous techniques is modified from hydrophobic to hydrophilic due to the attachment of polar groups, and thus they are easily soluble in many organic solvents. In real applications, many studies have proven that these techniques often produce CNT/polymer nanocomposites possessing higher mechanical and functional properties than those without functionalization.

3.3 Noncovalent Functionalization of CNTs

Functionalization of CNTs using covalent methods has shown promising results for providing useful functional groups onto the CNT surface. However, these methods have two major drawbacks, including creation of defects and an environmentally unfriendly process. The functionalization process inevitably creates a large number of defects on the sidewalls and/or the end tips of CNTs. In many extreme cases, CNTs are broken into smaller pieces, resulting in severe degradation of the mechanical properties of CNTs and disruption of their π electron system. The impact of disrupted π electrons is detrimental to the transport properties of CNTs because each defect site scatters the transportation of electrons or phonons. Concentrated acids or strong oxidants are often used for CNT functionalization, which are environmentally unfriendly. This means that development of new methods that are convenient to use, of low cost, and less damaging to CNTs is necessary to guarantee wider applications of CNTs in many emerging fields. From this viewpoint, functionalization based on noncovalent methods offers an alternative to overcome the drawbacks of covalent CNT functionalization.

According to the interactions between the CNTs and the guest molecules, the noncovalent functionalization methods can be further classified as the surfactant adsorption, polymer wrapping, and endohedral methods (Figure 3.3), and the principles of these methods are discussed in the following text.

3.3.1 Surfactant Adsorption

The physical adsorption of surfactant onto the CNT surface (Figure 3.3A) is designed to lower the surface tension of CNTs, effectively preventing the formation of aggregates. The surfactant-treated CNTs overcome the van der Waals attraction by electrostatic/steric repulsive forces. The efficiency of this method depends strongly on the type of surfactants used, medium chemistry, and the polymer matrix. In water-soluble polymers such as polyethylene glycol, ionic surfactants, such as Tween-80 (polyethylene oxide 20), were shown to promote dispersion, whereas in water-insoluble polymers like polypropylene and epoxy, CNT dispersion was promoted by nonionic

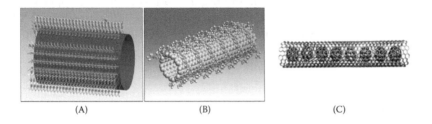

(A)   (B)   (C)

FIGURE 3.3
Schematics of CNT functionalization using noncovalent methods: (A) surfactant adsorption, (B) polymer wrapping, and (C) endohedral method. Adapted from Ma, P. C. 2008. Novel surface treatment, functionalization and hybridization of carbon nanotubes and their polymer-based composites. PhD dissertation. Hong Kong University of Science and Technology. 1–15.

surfactants, such as polyoxyethylene 8 lauryl, in organic solvents[40,41]. The role and effects of surfactants on CNT dispersion are discussed in Section 2.6, and thus further details are not included in this section.

3.3.2 Polymer Wrapping

The suspension of CNTs in the presence of polymers, such as poly(phenylene vinylene)[42], or polystyrene[43], results in wrapping of the polymer around the CNTs to form supermolecular complexes of CNTs (Figure 3.3B). It is a typical example of the noncovalent functionalization of CNTs, and this process is achieved through van der Waals interactions and π-π stacking between the CNTs and the polymer chains. The wrapping of CNTs with polymers that bear polar side chains, such as poly(vinyl pyrrolidine) (PVP), poly(styrene sulfonate) (PSS), poly(ethylene glycol) (PEG), poly(vinyl alcohol) (PVA), and so on, produced stable solutions of the corresponding CNT/polymer complexes in water[45,46]. The simulation results[44] suggested that the polymer backbone, not the side groups, provided a strong binding to the nanotubes, and it was energetically more favorable for the polymer to lie along the parallel axis instead of mapping onto the chirality of the underlying tube.

Wrapping of CNTs using conjugated polymers (containing π electrons) have attracted much interest due to improved π-π stacking between the CNTs and polymer matrix. Kang et al.[47] used an amphiphilic, linear conjugated polymer (poly[*p*-{2,5-bis(3-propoxysulfonicacidsodiumsalt)}phenylene]ethynylene, PPES) to functionalize SWCNTs in an aqueous phase under ultrasonication. The results showed that PPES was able to disentangle SWCNTs from their bundled forms at high mass percent conversion, producing suspensions that contained about 80% of the individually dispersed CNTs. The CNT concentration used in this study exceeded by an order of magnitude equivalent to the surfactant functionalized SWCNT benchmark. It was also found that PPES–wrapped SWCNTs formed a self-assembled helical super structure, and the experimentally observed PPES-SWCNT helical pitch of 13 ± 2 nm was in close agreement with 14 ± 1 nm predicted by a molecular dynamic simulation.

The properties of polymer–wrapped CNTs were markedly different from those of the individual components. For example, the complex consisting of SWCNTs and poly[(m-phenylenevinylene)-alt-(p-phenylenevinylene)] (PmPV) exhibited conductivity eight times higher than that of pure polymer, without any restriction on its luminescence properties. The SWCNT/ PmPV complex had a small average diameter of around 7.1 nm, suggesting that the bundles were mostly broken up on complex formation. The excellent optoelectronic properties of the SWCNT/PmPV complexes have been employed for manufacturing of photovoltaic devices[48,49]. In another study, Yi et al.[50] found that the intrinsic ability of PmPV in forming a helical conformation played an essential role in the separation of nanotubes.

Among about 15 tubes present in the pure SWCNT samples, PmPV was found to selectively pick up the tubes with chiralities of (11, 6), (11, 7), and (12, 6) with their diameters of 1.19, 1.25, and 1.24 nm, respectively. The SWCNTs of small diameters were held loosely by PmPV, and were gradually dropped out upon centrifugation. The suspension solution prepared from the SWCNT and PmPV was not permanently stable, and precipitation occurred after a few weeks. Irradiation in the UV-visible region exhibited a catalytic effect to shorten the precipitation time to hours. Those tubes that were held loosely by PmPV were quickly separated from the suspension during the irradiation process.

3.3.3 Endohedral Method

Another noncovalent method for CNT functionalization is the so-called endohedral method (Figure 3.3C)[1,3,23–25]. In this method, guest atoms, molecules, inorganic nanoparticles, such as C_{60}, Ag, Au, and Pt, or biomolecules, such as a protein like lactamase, are stored in the inner cavity of CNTs through the capillary effect. The incorporation of guest molecules is usually executed at defect sites localized at the ends or on the sidewall. Comparison of the catalytic activities of immobilized enzymes with those of the free species in the hydrolysis of penicillin showed that a significant amount of the inserted lactamase remained catalytically active, implying that no drastic conformational change had taken place. DNA can also be entrapped in the inner hollow channel of nanotubes by simple adsorption to form natural nano test tubes. Molecular dynamic simulations showed that both van der Waals and hydrophobic forces were important for the dynamic interactions of the components.

The combination of CNTs and guest molecules is particularly useful in catalysis, energy storage, nanotechnology, nanofluidics, and molecular scale devices because the incorporation process allows their individual properties to be integrated in the hybrid materials, thus creating unique properties that cannot be achieved with individual components acting alone.

To summarize the covalent and noncovalent functionalization techniques, the principles of these methods along with the corresponding advantages and disadvantages are presented in Table 3.1. It should be mentioned that there are many processing and material variables that need to be optimized if the techniques developed are to be fully exploited and thus to offer maximum benefits. While myriad studies have hitherto been directed toward modification of CNT surface characteristics, many of these studies have not provided optimized solutions. In addition, there have been concerns that excessive chemical processes often result in structural changes and severe damages to CNTs, which should be minimized in an effort to maintain the inherent mechanical and functional properties of CNTs. From this viewpoint, noncovalent functionalization methods are more favored than covalent techniques.

TABLE 3.1 Advantages and Disadvantages of Different Methods for CNT Functionalization.

Method		Principle	Damage to CNTs	Easy to Use	Interaction with Polymer Matrix*	Reagglomeration of CNTs in Matrix
Covalent Method(Physical functionalization)	Sidewall	Hybridization of C atoms from sp² to sp³	√	×	S	√
	Defect	Defect transformation and stabilization by organic groups or polymers	√	√	S	√
Noncovalent Method (Chemical functionalization)	Surfactant adsorption	Physical adsorption	×	√	W	×
	Polymer wrapping	van der Waals force, π-π stacking	×	√	V	×
	Endohedral method	Capillary effect, hydrophobic forces	×	×	W	√

Note: * S: Strong; W: Weak; V: Variable according to the miscibility between the matrix material and the polymer on CNTs.

3.4 CNT Interactions with Biomolecules

The unique properties of CNTs offer a wide range of opportunities and potentials for applications in biology and medicine[51-54]. For example, the rich electronic properties of CNTs have been explored for the development of highly sensitive and specific nanoscale biosensors. Promising results have been produced on the use of CNTs in various electroanalytical nanotube devices and as electromechanical actuators for artificial muscles. The optical absorption of CNTs in the near-infrared has been used for laser heating cancer therapy. The rapid progress of dispersion and functionalization techniques in recent years has helped to address many biocompatibility-related issues, opening up an even wider range of bioapplications such as drug delivery, bioconjugation, and recognition of CNTs with specific chiralities. Therefore, the development of controllable interactions between CNTs and biomolecules represents another solid step toward the widespread use of CNTs in biological and biomedical fields.

3.4.1 Interaction with DNA

Deoxyribonucleic acid (DNA) is a nucleic acid that contains the genetic instructions used in the development and functioning of all known living organisms and some viruses. The main role of DNA molecules is the long-term storage of information. DNA is often compared to a set of blueprints, a recipe, or a code, since it contains the instructions needed to construct other components of cells, such as proteins and ribonucleic acid (RNA). Chemically, DNA consists of two long polymers of simple units called *nucleotides*, with backbones made of sugars and phosphate groups joined by ester bonds. These two strands run in opposite directions to each other and are therefore antiparallel. DNAs can be attached to CNTs via physical adsorption or covalent bonding by coupling reactions either directly or through a bifunctional linker[54]. The best stability, accessibility, and selectivity can be achieved through covalent bonding because of its capability to control the location of the biomolecules, improved stability, accessibility, selectivity, and reduced leaching[55]. In order to covalently attach the molecules to nanotubes, the first requirement is the formation of functional groups on the CNTs. The carboxylic acid group is often the best choice because it can undergo a variety of reactions and is easily formed on CNTs via oxidizing treatment. For example, Williams et al.[56] used a bifunctional linker approach to functionalize sidewalls of CNTs in several steps, starting with an acid treatment of CNTs to generate carboxylic groups, which were then esterified by N-hydroxysuccinimide, followed by coupling with peptide nucleic acid (PNA, an uncharged DNA analogue) and hybridizing these macromolecular wires with complementary DNA. These step-by-step processes are schematically shown in Figure 3.4A[56]. The study provided a new, versatile means of incorporating SWCNTs into large electronic devices by a recognition-based assembly, and of using SWCNTs as probes in biological systems by sequence-specific attachment.

The mechanisms behind the physical interactions between the CNTs and DNA is not fully understood, though a number of driving forces have been proposed[54], such as van der Waals and π-π stacking, entropy loss due to the confinement of the DNA backbone, and electronic interactions between DNA and CNTs. Heller et al.[57] investigated the DNA-nanotube conjugates and the associated conformational change of DNA on the CNT surface. It was found that the double-stranded DNA wrapped on SWCNTs underwent conformational transition from the right-handed B form to the left-handed Z form upon the addition of divalent cations such as Hg^{2+} (Figure 3.4B). Thus, these DNA-wrapped CNTs were used to detect ions in blood, tissues, and living mammalian cells.

Functionalization of CNTs with DNA not only affects the aqueous solubility of CNTs, but also allows a more precise control of the interfacial properties of CNTs. For example, Lu and coworkers[58] developed a method to adsorb poly-T DNA onto a CNT surface via a π–π interaction and employed this complex as ultrathin dielectrics for nanotube electronics. The functionalization made it possible to fabricate SWCNT field effect transistors free of gate-leakage for potentially ultimate performance. The same research group also grafted short interfering ribonucleic acid (siRNA) onto the CNT surface, where the thiol-modified

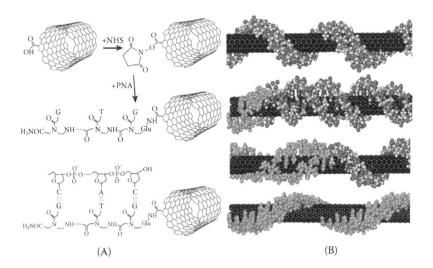

(A)   (B)

FIGURE 3.4

(A) Covalent attachment of DNA to CNTs via esterification and amidation; adapted from Williams, K. A., et al. 2002. *Nature* 420: 761, and (B) DNA undergoing a conformational transition from the B form (top) to the Z form (bottom) on a CNT; adapted from Heller, D. A., et al. 2006. *Science* 311: 508.

siRNA cargo molecules were linked to aqueous solubilized SWCNTs, which were previously functionalized by an amine-terminated surfactant[59]. According to the results obtained from vitro delivery experiments, these siRNA-SWCNT conjugates could effectively silence (i.e., 50–60% knockdown) both the CD4 receptor and the CXCR4 co-receptor that are required for HIV to enter human T cells and infection, confirming great potential of using DNA-functionalized CNTs in medical applications.

The incorporation of DNA molecules onto CNTs endows the new hybrid material with biological activities. Luciferase DNA containing the T7 promoter sequence was adsorbed onto SWCNT buckypaper and about 36% of the bound DNA remained functional as compared to the native DNA (solution-based) in an RNA polymerase-catalyzed transcription–translation reaction[60]. The interaction between the DNA and CNTs was also generally stable, allowing the separation of the dispersed CNTs into well-defined subpopulations[54]. Wrapping of CNTs by single-stranded DNA (ssDNA) was found to be sequence dependent, and a specific sequence of the ssDNA could self-assemble into a helical structure around the individual nanotubes. The electrostatics of the DNA-CNT hybrid depended on the diameter and electronic properties of CNTs, enabling nanotube separation by anion exchange chromatography. Optical absorption and Raman spectroscopy revealed that early fractions were enriched in the CNTs with small diameters and metallic characteristics, whereas late fractions were favored in the CNTs with large diameters and semiconducting characteristics[61].

3.4.2 Interaction with Proteins

Proteins, also known as polypeptides, are organic compounds consisting of amino acids arranged in a linear chain and folded into a globular form. The amino acids in a protein are joined together by peptide bonds between the carboxyl and amino groups of adjacent amino acid residues. Proteins can interact with CNTs via either physical adsorption or immobilization where the reactions are controlled easily by utilizing the functional groups, such as the amine, amide, and carboxylic groups[54]. Chen et al.[62] reported a general approach to immobilize various proteins onto the CNT surface by a noncovalent method. This process involved the nucleophilic substitution of physically adsorbed groups like N-hydroxysuccinimide on CNTs by an amine group on the protein, followed by the formation of an amide bond. It was shown that this technique enabled the immobilization of a wide range of proteins on the sidewalls of CNTs with a high degree of control and specificity.

Nepal et al.[63] demonstrated that common proteins, such as lysozyme, histone, hemoglobin, myoglobin, ovalbumin, bovine serum albumin, trypsin, and glucose oxidase, were good dispersing agents for SWCNTs and that the ability to disperse by these proteins depended on various factors including their primary structures and pH. Proteins with more basic residues, like histone and lysozyme, were more effective in the dispersion of SWCNTs, possibly due to the π–π interactions. Regarding the pH effect on functionalization of SWCNTs with hemoglobin, either a low (acidic) or high (basic) pH was required for enhanced solubilization.

Proteins have been widely used to stabilize the formation of metal nanoparticles (NPs); thus it is possible to anchor metal NPs onto the CNT surface via the interactions between CNTs and protein. Wei et al.[64] functionalized MWCNTs with an aromatic chemical, 1-pyrenebutyric acid N-hydroxysuccinimide (PANHS) ester (Figure 3.5A), and produced four protein- (fibrinogen, c-globulin, hemoglobin, and fibronectin) protected gold NPs by a simple strategy of reducing gold ions (Figure 3.5B). When the above two solutions were mixed, the proteins mediated the assembly of protein-protected Au-NPs on the functionalized MWCNTs, resulting in the formation of Au-NP/CNT nanohybrids (C and D in Figure 3.5). This study offered a new way to prepare well-defined metal NP-coated CNTs, which showed potential applications of protein-protected metallic NPs for preparing novel nanomaterials.

Another application of CNT-protein complex is the sorting of CNTs with different chiralities. In a study by Poenitzsch et al.[65] on the dispersion of SWCNTs with amphiphilic helical peptides, the effectiveness of the peptides was found to be dependent on the electron density of the aromatic residue on the hydrophobic face. The proteins exhibited high selectivity toward the metallic CNTs, thus the CNT-protein complex offered a way to separate metallic CNTs from the as-produced nanotubes, which is generally a mixture of metallic and semiconducting CNTs.

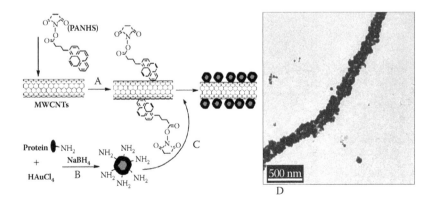

FIGURE 3.5
Model representation of protein-mediated formation of CNT-Au-NP hybrids: (A) functionalization of MWCNTs with a PANHS, (B) preparation of protein-protected Au-NPs, (C) self-assembly of protein-protected Au-NPs on MWCNTs by nucleophilic substitution of NHS group with primary and secondary amines on the surface of the protein, and (D) typical TEM image of MWCNT-protein-Au-NP hybrids. Adapted from Wei. G., et al. 2010. *Carbon* 48: 645.

3.4.3 Interaction with Living Cells

Many studies have revealed that both SWCNTs and MWCNTs could be internalized by a variety of living cells to deliver therapeutic and diagnostic small molecules and macromolecules to the cells[66]. The possible mechanism behind this interaction involves cellular binding, internalization, and intracellular trafficking. These investigations using CNTs have instigated vigorous debates in the rapidly developing field of CNT biotechnology. One of the prerequisites for exploring the interaction of CNTs with living cells is to adequately overcome the issue of CNT compatibility with the aqueous biological environment. Pantarotto et al.[67] found that CNTs were able to interact with plasma membranes and cross into the cytoplasm without having to be engulfed in a cellular compartment to facilitate intracellular transport. Model nanotube structures were also proposed to interact with lipid bilayers via a diffusion process directly through the biomembrane, as illustrated in Figure 3.6A. Spontaneous transmembrane penetration of CNTs via the flipping of membrane lipids is independent of energy, receptor, coat, or lipid raft interactions, and is potentially relevant to all cell types. Lopez et al.[68] also described the spontaneous diffusion of functionalized CNTs with hydrophilic termini through the lipid bilayer membranes using molecular dynamic simulations, further confirming the experimental results.

Two high-resolution images of amino-functionalized MWCNTs taken during initial contact, interaction (Figure 3.6B), and penetration (Figure 3.6C) of mammalian cells are presented in Figure 3.6. CNTs were found oriented perpendicular to the plasma membrane of the cells during the cellular

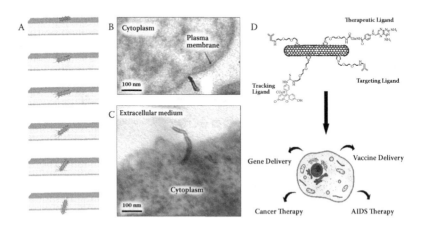

FIGURE 3.6

Interaction between CNTs and cells and their potential applications: (A) schematic of a CNT crossing the plasma membrane, (B) TEM image of CNTs interacting with the plasma membrane of living cells, (C) TEM image of CNTs crossing the plasma membrane of living cells, and (D) typical bioapplications of CNTs as novel delivery systems in a cell. Adapted from Lacerda, L., et al. 2007. *Nano Today* 2: 38.

internalization. The selection of cell types was considered important because nonprofessional phagocytosis could also contribute to the cellular internalization of CNTs. There were various reports of CNT internalization using multiple cell types, such as fibroblasts, epithelial, cancer cells, phagocytes, bacteria, and fungi, under different experimental conditions[66]. The effect of functional groups present on the CNT surface has also been investigated using techniques such as confocal microscopy and fluorescence-activated cell sorting. It was also confirmed that CNTs were able to pierce or penetrate into the plasma membrane by a process independent of energy, regardless of cell type or characteristics of the functional groups attached to the CNTs.

In addition, the hypothesis of CNTs acting as "nanoneedles" with the cell has been experimentally reproduced for two different types of CNTs: (1) block copolymer-wrapped MWCNTs using microglia cells[69], and (2) oxidized, water-soluble CNTs interacting with *E. coli* in the presence of microwaves[70]. These studies further confirmed that the nanoneedle mechanisms other than classical endocytosis were contributing to the high levels of cellular interactions with CNTs. The interesting properties of water-dispersible, individualized CNTs can be used in biomedical applications, for example, as novel carrier systems for therapeutics and diagnostics. These applications are based on the fact that (1) CNTs can be internalized by a wide range of cell types, and (2) their high surface area can potentially act as a template for cargo molecules such as peptides, proteins, nucleic acids, and drugs. CNTs have been described as delivery systems mainly in proof-of-principle studies for a variety of different biomedical applications ranging from gene delivery to cancer therapy, as shown in Figure 3.6D. A comprehensive review of biomedical applications of CNTs can be found in reference[66].

3.4.4 Interaction with Carbohydrates

Carbohydrates are highly hydrophilic biomolecules and play numerous roles in the immune system, fertilization, pathogenesis, blood clotting, and development of living things. For example, polysaccharides serve as the storage of energy (e.g., starch and glycogen) and as structural components (e.g., cellulose in plants and chitin in arthropods). The 5-carbon monosaccharide ribose is an important component of coenzymes and the backbone of the genetic molecule, known as DNA.

To make CNT soluble in aqueous media, the possibility of decorating the graphitic surface with carbohydrate macromolecules has been explored. In the work of Bandyopadhyaya and coworkers[71], it was shown that CNTs could be efficiently untangled and dispersed in an aqueous solution of gum arabic by nonspecific physical adsorption. Gum arabic is a highly branched arabinogalactan polysaccharide. Star et al.[72] studied the complexation of nanotubes with starch and, in particular, its linear component amylose. The polysaccharide consisted of glucopyranose units and adopted a helical conformation in water, forming inclusion complexes with various substances. The initial experiment revealed that the CNTs were not soluble in an aqueous solution of starch, but rather were soluble in a solution of a starch-iodine complex. This observation suggested that the preorganization of amylose in a helical conformation through complexation with iodine was critical for a single tube to enter the cavity of the helix.

Using dimethyl sulfoxide–water mixtures, Kim et al.[73] reported the solubility of CNTs with amylase where the polysaccharide adopted an interrupted loose helix structure. It was claimed that the helical state of amylose was not a prerequisite for nanotube encapsulation. In addition, the dispersion capabilities of other amylose homologues, pullulan and carboxymethyl amylase were also studied, confirming that these substances had lower solubility than amylose.

The CNT-displayed carbohydrates not only impart significant aqueous solubility and biocompatibility, but also offer bioactivities that are apparently not available with other display platforms, such as polymeric nanoparticles. For example, Sun and coworkers[54,74] demonstrated that the monosaccharide (e.g., galactose or mannose)–functionalized SWCNTs could effectively bind and aggregate anthrax spores in the presence of calcium ions with a 97.7% reduction. Interestingly, however, the polymeric nanoparticles—polystyrene beads of around 120 nm in diameter—functionalized with the same sugars exhibited no similar binding or aggregation of the spores under the same experimental conditions, suggesting the uniqueness of SWCNT as a linear and semiflexible scaffold for multivalent displaying of the monosaccharide. The Ca^{2+}-mediated binding and aggregation of anthrax spores were found to be reversible, and complete deaggregation of spores occurred when free Ca^{2+} were removed by adding the chelating agent, ethylene diamine tetraacetic acid. Therefore, from the mechanistic viewpoint, the binding was a result of divalent cation-mediated carbohydrate–carbohydrate interactions between the SWCNT-displayed multivalent monosaccharides and the sugar moieties on the spore surface.

3.5 CNT Functionalization in Different Phases

3.5.1 CNT Functionalization in the Liquid Phase

Many research efforts have been directed toward dispersion and functionalization of CNTs for producing CNT/polymer nanocomposites. Most of these studies, whether covalent or noncovalent functionalization, and mechanical or physical dispersion, are often accomplished in liquid media. The liquid media include water and organic solvents. The use of liquid media is widespread because:

1. For the functionalization of CNTs, one of the prerequisites is to achieve good CNT dispersion, which guarantees an effective "contact" between the CNTs and the functioning agent. Better CNT dispersion can be achieved in some liquids using a certain dispersion technique than others, according to the discussion presented in Section 2.5.

2. Functionalization of CNTs in a liquid results in an easy separation and purification after reaction because they are not truly dissolved in most liquids but are conveniently isolated from liquids either by filtration or by centrifugation.

3. A wide variety of liquids are available for CNT functionalization. For covalent functionalization of CNTs, some reactions can be accomplished only in certain liquid media. In noncovalent methods, such as surfactant adsorption and polymer wrapping, liquids play an important role in stabilizing and diluting these modifying molecules, thus making it possible to lower the cost of the functionalization process.

4. In many operations, dispersion and functionalization of CNTs as well as following mixing with monomer/polymer and polymerization take place at the same time in one step without separation, thus providing a convenient way to fabricate CNT/polymer nanocomposites.

3.5.2 CNT Functionalization in the Solid Phase

In contrast to the myriad reports on CNT functionalization in liquid media, there are only a few papers describing processes in a solid phase. A typical example is to functionalize CNTs based on ball milling, which is also known as a mechanochemical method. Ball milling is a type of process to grind materials into extremely fine powder for use in paints, pyrotechnics, and ceramics. The principle of this technique is introduced in Section 2.3.3. The effects of ball milling on the CNT structure and dispersion have been well documented, however, little is known concerning their effects on the functionalization of CNTs.

Ma et al.[75,76] developed a simple but efficient mechanochemical method to in-situ functionalize MWCNTs based on ball milling. The elemental compositions and the functional groups introduced on the CNT surface were evaluated after ball milling in the presence of ammonium bicarbonate, NH_4HCO_4. Figure 3.7A summarizes the x-ray photoelectron spectroscopy (XPS) results which illustrate the variations of oxygen-to-carbon and nitrogen-to-carbon ratios of CNTs as a function of milling time. The nitrogen content increased continuously as the ball milling time increased, suggesting the attachment of nitrogen compounds onto the CNT surface via covalent bonding and/or physical adsorption. The authors further illustrated the mechanism behind the successful attachment of nitrogen compounds on the CNT surface (Figure 3.7B): during milling, NH_4HCO_4 was decomposed into NH_3 gas,

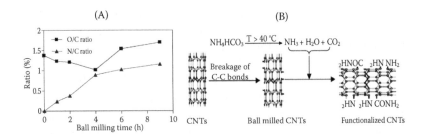

FIGURE 3.7
(A) Variations of oxygen-to-carbon and nitrogen-to-carbon ratios of CNTs as a function of milling time, and (B) schematics showing attachment of amine and amide groups on CNTs due to ball milling. Adapted from Ma, P. C., et al. 2008. *Chem. Phys. Lett.* 458: 166 and Ma, P. C., et al. 2009. *J. Nanosci. Nanotechnol.* 9: 749.

water, and CO_2 while the milling process also introduced defects and breaks the –C–C– bonds of the CNT surface, which was further promoted by the presence of NH_4HCO_4. These in turn allowed the amine and amide groups to form covalent bonds with the broken –C–C– bonds.

The ball milling of CNTs in reactive atmospheres, such as gases containing chemical functional groups such as amines, amide, thiols, and mercaptans, was shown to produce short CNTs possessing functional groups in high quantity[77]. In an analogous strategy, SWCNTs were reacted with potassium hydroxide through a simple solid-phase milling technique[78]. The nanotube surface was covered with hydroxyl groups, and the derivative displayed an improved solubility in water (up to 3 mg/mL). Using the same approach, fullerene-C_{60} could also be attached to the graphitic network of nanotubes[79]. This was confirmed by the featureless absorption spectrum of SWCNT-C_{60} and the increased intensity of the disordered mode in the Raman spectrum.

3.5.3 CNT Functionalization in the Gas Phase

One of the unique advantages of CNT functionalization in a gas phase is that it is an easy and environmentally friendly process because the tedious washing and separation processes that are commonly required in liquid reactions can be totally avoided. Currently, *ultraviolet/ozone (UV/O₃)* and *plasma* are two typical examples of gas-phase functionalization.

UV/O₃ has been widely adopted in many areas, including aquaculture applications for disinfection to improve water quality, cleaning and modification of polymer substrates, and functionalization of carbon materials. A comprehensive review of this topic has recently been presented by Sham and coworkers[80]. Basically, UV/O₃ treatment is a photosensitized oxidation process where the molecules on either the substrate surface or the organic contaminants are excited and dissociated by the absorption of short-wavelength UV radiation. Once exposed to UV radiation, targeted materials, such as CNTs, will typically react with the atomic oxygen arising from the continuous dissociations of oxygen and ozone molecules according to the following mechanisms (Figure 3.8)[80–83]:

1. Changes of molecular oxygen to excited-state molecular oxygen upon UV radiation at a wavelength less than 240 nm.
2. Dissociation of the excited-state oxygen into two ground-state oxygen atoms.
3. Generation of ozone from the ground-state oxygen atoms and molecular oxygen.
4. Photolysis of ozone upon UV radiation into atomic oxygen and molecular oxygen, where the atomic oxygen is extremely reactive and short-lived, and tends to react with the exiting gaseous species, like molecular oxygen and ozone, within the UV/O$_3$ chamber. Hydroxyl radicals are also produced in the presence of water vapor. Hydroxyl and carboxyl groups are likely found on CNTs after the treatment.

There have been several studies that reported the modification of CNTs based on the UV/O$_3$ technique. In fact, the UV/O$_3$ treatment was often used to modify the CNT surface chemistry in combination with other functionalization techniques, such as amino treatment[84] and silanization[85]. This is because the UV/O$_3$ technique alone can only introduce hydroxyl, carboxyl, and carbonyl functional groups, and is not sufficient to form strong covalent bonds with polymer resins. For example, a scheme has been proposed to modify the

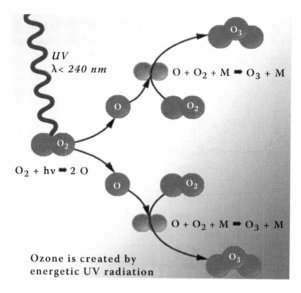

FIGURE 3.8
Schematic showing the generation of O$_3$ upon UV irradiation. Adapted from The Ozone Hole, http://www.theozonehole.com/ozonecreation.htm (accessed December 2009).

CNT surface by incorporating both UV/O$_3$ and a triethylenetetramine (TETA) solution[84]. The UV/O$_3$ treatment was aimed at promoting oxygen-containing functional groups, and upon combination with the TETA treatment, it allowed amine-functional groups to be introduced on the CNT surface, as confirmed by the FT-IR spectra shown in Figure 3.9. The carboxylic carbon consisting of ester and acid steadily increased with increasing UV/O$_3$ exposure time, whereas the contents of other functional groups, such as ether (C–O–C) and hydroxyl (C–OH) groups, remained basically unchanged (Figure 3.9A). Upon treatment with TETA following the UV/O3 exposure for 30 min, new peaks appeared at wave numbers at 2926 and 2854 cm^{-1} corresponding to N-H stretching vibrations, and the peaks at 3100~3400 cm^{-1} and 1660 cm^{-1} were related to amine vibrations (Figure 3.9B). The presence of these functional groups was proven to be beneficial to dispersing CNTs in the epoxy matrix by transferring the hydrophobic nature of pristine CNTs to a more hydrophilic character.

Figure 3.10 shows the effects of UV/O$_3$ treatment on changes in surface energy of CNT films, which originate from the goniometry of CNT film with different probing liquids. The surface energy has two components, namely the polar and dispersive (or nonpolar) components. It is worth noting that after UV/O$_3$ treatment, the polar component increased considerably, while the corresponding dispersive component remained unchanged. The enhanced polar component after the UV/O$_3$ treatment was a reflection of newly incorporated functional groups on the CNT surface, such as hydroxyl, carboxyl, and carbonyl groups. However, a further increase in treatment time beyond 30 minutes did not have a significant effect on surface energy, indicating saturation of these functional groups[84].

While the chemical functionalization techniques in liquid media, such as HNO$_3$, H$_2$SO$_4$, or a mixture of them, or with oxidants such as KMnO$_4$, can

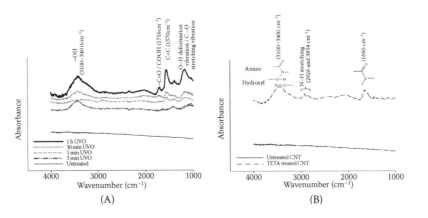

FIGURE 3.9
(A) FT-IR spectra of MWCNTs after UV/O$_3$ treatment with different exposure times, and (B) after UV/O$_3$ treatment for 30 minutes followed by amino-functionalization with TETA. Adapted from Sham, M. L., et al. 2006. *Carbon* 44: 768.

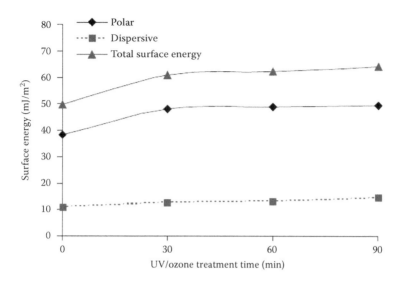

FIGURE 3.10
Variation of surface energy of modified CNTs as a function of UV/O₃ treatment time. Adapted from Sham, M. L., et al. 2006. *Carbon* 44: 768.

create covalent bonds between the treated CNTs and polymer resins, many of these methods are detrimental to the inherent mechanical properties of CNTs because they inevitably generate defects on the CNT surface as a result of breakage of carbon–carbon bonds. A major advantage of UV/O₃ treatments is that the reactions of UV/O₃ are milder than the above oxidative methods in liquid media[84]. Furthermore, the effectiveness of the UV/O₃ treatment can be enhanced by optimizing the processing parameters, such as exposure time, lamp power, lamp–subject distance, temperature, feed gases, and so on.

Besides CNTs, UV/O₃ treatment has also been successfully employed to functionalize graphite nanoplatelets (GNPs), with the intention of enhancing the interfacial adhesion between the GNP and polymer resins[86,87]. The UV/O₃-exposed GNPs presented a rougher surface with sharper boundaries between the basal planes than those without treatment. The loosely bound materials and organic contaminants on the GNP were removed through the etching process of the UV/O₃ treatment. The treatment introduced a significant amount of oxygen-containing functional groups on the GNP surface, resulting in a much higher storage modulus below the glass transition temperature for the nanocomposite containing UV/O₃-treated GNPs. A similar improvement on electrical and thermal conductivities of GNP/epoxy nanocomposites were also noted because the cleaned GNP surface would eliminate the unbounded microvoids at the GNP polymer matrix interface, which might act as insulating pockets with high–contact resistance.

An alternative solid-gas approach to functionalize CNTs involves radiofrequency plasma activation. *Plasma* is a gas in which a certain portion of gas molecules are ionized. The presence of charge carriers makes the plasma electrically conductive so that it responds strongly to electromagnetic fields. Plasma therefore has properties quite unlike those of solids, liquids, or gases and is considered to be a distinct state of matter. Like gas, plasma does not have a definite shape or a definite volume unless enclosed in a container; unlike gas, however, it may form structures such as filaments, beams, and double layers under the influence of a magnetic field.

Plasma is commonly used for the synthesis of CNTs with a specific alignment. The first report of CNT functionalization using plasma studied the effects of three different oxidative treatments, namely air oxidation, oxygen plasma oxidation, and acid oxidation on the surface properties of MWCNTs[88], showing that these oxidative treatments affected the valence bands of CNTs and increased their work functions. XPS measurements suggested that the treatments using air and oxygen plasma preferentially introduced hydroxyl and carbonyl groups, while the liquid-phase treatment induced carboxylic acid groups on the surface of MWCNTs. These surface groups disrupted the π-conjugation and introduced surface dipole moments, leading to higher work functions by up to 5.1 eV. The information on the work function of MWCNTs is of importance to the development of electronic or optoelectronic applications.

To understand the mechanisms of CNT functionalization using plasma, Chen et al.[89] employed a microwave-excited NH_3/Ar plasma to functionalize MWCNTs, and investigated the effects of several parameters, such as plasma flow rate, treatment time, microwave power, and bias voltage on the elemental compositions and structural properties of CNTs. The results showed that NH radicals were created in the plasma zone and other fractionation products, for example, N* and H* interacted with the CNT surface, finally eliciting surface-bound NH radicals. The NH radicals are highly reactive and interact with the nonactivated and activated sites on the MWCNT surface, resulting in the formation of amino groups. The concentration of primary amino groups was affected by the plasma flow rate, microwave power, and bias voltage. The pure NH_3 gave rise to unstable plasma, which in turn adversely affected the properties of functionalized CNTs and the reproducibility of the results. In contrast, the Ar addition to the system made plasma stable, allowing the properties of modified CNTs reproducible.

3.6 Effects of Functionalization on the Properties of CNTs

3.6.1 Dispersability and Wettability of CNTs

One of the major objectives of functionalizing CNTs is to improve their dispersion both in solvent and in polymer matrices, which indeed has been confirmed by myriad publications. As noted in Section 2.4, a proper evaluation of CNT dispersion involves examinations on different scales. The macroscopic evaluation can be done by comparing the suspension stability of CNT solutions at different time durations, whereas TEM or SEM is a powerful tool to study CNT dispersion on the micro- and nanoscales. Ma et al.[90] functionalized MWCNTs using an organic silane, and compared the dispersion behaviors of CNTs before and after functionalization. Figure 3.11 shows the suspending states of the MWCNTs in ethanol with a fixed CNT concentration of 0.5 mg/ml after 5 minutes of dispersion under ultrasonication (US). The comparison of these images indicated that the suspension stability of the pristine MWCNTs (A in Figure 3.11) was poor and they deposited totally in ethanol due to the agglomeration and poor hydrogen-bonding ability. The silane functionalized CNTs (E in Figure 3.11) showed much better stability than the pristine ones both in the short and long terms. It was concluded that the silane functional groups on the CNT surface enhanced their hydrogen-bonding abilities and converted the CNT surface from the hydrophobic state to the hydrophilic one.

Figure 3.12 shows typical TEM images of the MWCNTs functionalized using the above method. The pristine MWCNTs were severely agglomerated (Figure 3.12A) whereas the modified MWCNTs (Figure 3.12B) showed much improved dispersion with individual MWCNTs detached loosely without significant changes in their lengths. These results are in agreement with the observations from Figure 3.11 in that the chemically modified MWCNTs exhibited improved suspension stability in organic solvent.

The change of surface properties of CNTs resulting from functionalization will obviously affect their wettability. Ma et al.[91] studied the effects of amino functionalization (ethylene diamine) on the dispersion and interfacial interaction of CNTs with solvents and polymer resins. The static contact angles of water and ethylene glycol on CNT films were measured, and the results are summarized in Table 3.2. The hydrophilicity increased substantially after

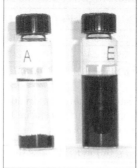

FIGURE 3.11
Suspension stability of CNTs in ethanol before (A) and after (E) silane functionalization. Adapted from Ma, P. C., et al. 2006. *Carbon* 44: 3232.

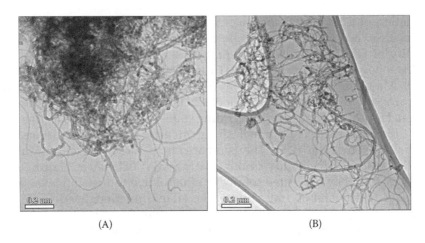

(A)   (B)

FIGURE 3.12
Dispersion of CNTs (A) before and (B) after silane functionalization. Adapted from Ma, P. C., et al. 2006. *Carbon* 44: 3232.

amino-functionalization as indicated by the significant reduction in contact angle from 72.0_0 to 43.0_0 when the polar probing liquid, that is, water, was used. A similar trend was also observed when the contact angles were measured using ethylene glycol as a probing liquid. Table 3.3 compares the surface energies of CNTs calculated based on Young's equation and the harmonic mean method[91,92]. The total surface energy of pristine CNTs was 42.1 mJ/m^2, which is similar to that of graphite (40.3 mJ/m^2)[93], and this result is also in agreement with a theoretical estimate of 40–80 mJ/m^2 for typical CNTs[94,95]. After functionalization, the surface energy of CNTs increased by more than 40%, and the increase arising mainly from the increase in polar components (γ_S^p, from 14.3 to 30.9 mJ/m^2), while the dispersive component of surface energy (γ_S^d) was essentially unchanged. The polar components of surface

TABLE 3.2   Contact Angles of Various Probing Liquids on a CNT Film Substrate.

	Water (°)	Ethylene glycol (°)	Epon 828 (°)
Pristine CNTs	72.0 ± 3.1	31.8 ± 2.4	74.1 ± 2.6
Amino CNTs	43.0 ± 2.1	21.8 ± 3.7	61.5 ± 3.2

Source: Adapted from Ma, P. C., et al. 2010. *Carbon* 48: 1824.

TABLE 3.3   Total Surface Energy (γ_S, Including Dispersive (γ_S^d) and Polar (γ_S^p) Components) and Interfacial Energies (γ_{LS}) of CNTs with Different Surface Conditions.

CNTs	γ_S (mJ/m^2)	γ_S^d (mJ/m^2)	γ_S^p (mJ/m^2)	$\gamma_{LS}{}^*$ (mJ/m^2)
Pristine CNTs	42.1	27.8	14.3	29.5
Amino CNTs	60.8	29.9	30.9	38.9

Note: * Interfacial energy of Epon 828/CNT interface.
Source: Adapted from Ma, P. C., et al. 2010. *Carbon* 48: 1824.

energy includes dipole–dipole, dipole-induced dipole, and hydrogen-bonding interactions[92], thus this enhancement is a reflection of newly incorporated functional groups on the CNT surface.

The contact angles of a typical epoxy monomer resin, Epon 828, on the CNT films were also measured in an effort to evaluate the interfacial interactions. As shown in Table 3.2, the contact angle of CNTs decreased from 74.1_o to 61.5_o after amino functionalization, suggesting enhanced wettability between the CNTs and epoxy. The corresponding interfacial energies (γ_{LS}) were calculated based on the reported surface energy of epoxy (Epon 828), $\gamma_L \sim 46.0$ mJ/m^2 [96]: $\gamma_{LS} = 29.5$ and 38.9 mJ/m^2 were obtained for the pristine CNTs and amino CNTs, respectively. The higher the interfacial energy, the larger the interaction between the solid substrate and liquid. Thus, the above observation further confirmed the effectiveness of amino-functionalization in enhancing the interfacial interactions between the CNTs and the polymer matrix.

3.6.2 Mechanical Properties

CNTs exhibit excellent mechanical properties, and they are regarded as the strongest and stiffest fibers ever known. Contrary to the extensive efforts made to measure the inherent mechanical properties of pristine CNTs, very little information is available so far regarding the effect of functionalization on the change in mechanical properties of CNTs. Possible reasons for this observation are as follows:

1. Difficulties associated with the sample preparation and separation of individual CNTs for testing. The determination of CNT mechanical properties involves the manipulation of samples in nanometer scale (i.e., based on TEM, atomic force microscopy [AFM], or scanning probe microscopy [SPM]). It is not trivial to grip and load the CNT samples, and to obtain repeatable data with sufficiently high resolution.

2. A very large variety of methods developed for CNT functionalization. Even for the same kind of CNTs, many different methods can be employed for functionalization (physical or chemical) for particular end applications, causing largely different effects on bonding states, surface properties, optical, electrical, and thermal properties of CNTs. Therefore, evaluation of the mechanical properties of functionalized CNTs requires understanding of other associated properties affected by the functionalization process.

3. There has also been a concern with respect to the degree of CNT functionalization. Excessive functionalization of CNTs severely damages them and shortens them into smaller fragments with lower aspect ratios. However, the quantitative evaluation of the degree of CNT functionalization is difficult, as is the correlation between the degree of functionalization and the change in mechanical properties.

Theoretical studies have been made concerning the effects of function-alization on the mechanical properties of CNTs. For example, Garg et al.[97] used classical molecular dynamics simulations to investigate the effect of covalent functionalization on the mechanical properties of CNTs. The results showed that the introduction of sp³-hybridized carbon defects due to chemical functionalization degraded the strength of CNTs by an aver-age 15% of the original value. Specifically, the buckling force for the (15, 15) SWCNTs decreased by about 19.4%, suggesting that the functionalized CNTs were less stiff in the direction of the tubular axis than the regular unfunctionalized CNTs, and thus are expected to deform more readily in a composite. When small CNTs with radii around 0.3–0.4 nm were deformed, the strain arising from the external force coupled with the inherent strain caused the chemically attached groups to dissociate from the CNT walls. Hence, while calculations and measurements have shown that nanotubes with smaller diameters are stiffer than those with larger diameters, their diameters should not be too small to avoid destabilization of the residual strains in the functionalized CNTs.

In another study, Zhang et al.[98] studied the effect of hydrogenization on the mechanical properties of SWCNTs using atomistic simulations. The results showed that the elastic modulus of CNTs gradually decreased with an increasing percentage of C–H bonds due to functionalization. However, both the strength and ductility dropped sharply at a low degree of function-alization, reflecting the sensitivity of these properties on C–H bonds. The cluster C–H bonds with two rings were mainly responsible for the signifi-cant reductions in strength and ductility of CNTs. The effect of carboniza-tion was essentially the same as that of hydrogenization.

The above studies revealed detrimental effects of chemical function-alization on important mechanical properties of CNTs. However, some mechanical properties of SWCNTs, including the critical bending curva-ture and the critical bending moment, were improved after functionaliza-tion[99]. The results from a molecular mechanics simulation showed that the functionalized SWCNTs were able to sustain higher external bending loads and larger bending deformation than pristine SWCNTs, and there existed an optimum degree of functionalization at which the critical curvature reached the maximum value. The critical curvature varied little beyond the optimum value. Taking into account the functional groups containing hydrogen atoms, it is revealed that the optimum degree of functionaliza-tion was about 24% regardless of the chiralities, diameters, and aspect ratios of SWCNTs. However, the bending rigidity of the SWCNT decreased with an increasing degree of functionalization. Although the previously stated understanding does not completely establish the correlation between the mechanical properties of CNTs and the corresponding process for func-tionalization, these studies shed light on the extent to which CNTs should be functionalized with respect to practical CNT reinforced nanocompos-ites designs.

3.6.3 Electrical/Electronic Properties

It is recognized that functionalization is one of the most convenient methods to modify the electrical/electronic properties of CNTs for many practical applications. Covalent functionalization disrupts the π-bonding system and breaks the translational symmetry of CNTs by introducing saturated sp^3 carbon atoms. As a result, electronic and transport properties of CNTs are significantly altered. There have been theoretical and experimental studies in this area. For example, Park et al.[100] presented the transport properties of covalently functionalized metallic SWCNTs based on a numerical study. The systematic dependence of CNT conductivity with monovalent and divalent sidewall additions was studied on the addend concentration or degree of functionalization. For monovalent bonding, the addend-induced impurity state resided near the Fermi level, which acted as a strong scattering center, adversely affecting the ballistic conducting properties. The conductance of CNTs decreased rapidly with addend concentration and approached zero at around 25% addend-to-C ratio. In contrast, the divalent addend had a negligible effect on conductance near the Fermi level. When increasing the addend concentration, the conductance was reduced gradually, and even at 25% concentration, the conductance remained at more than 50% of that of a perfect nanotube.

Zhao et al.[101] studied noncovalently functionalized SWCNTs that were adsorbed with various gas molecules, such as NO_2, O_2, NH_3, N_2, CO_2, CH_4, H_2O, H_2, and Ar, using the first principle methods. The equilibrium position, adsorption energy, charge transfer, and electronic band structures were obtained for SWCNTs. The results showed that most molecules adsorbed weakly onto SWCNTs and served as either charge donors or acceptors to nanotubes. The gas adsorption on the interstitial and groove sites of CNT bundles was stronger than that on individual nanotubes. The electronic properties of SWCNTs were found to be sensitive to the adsorption of certain gases, such as NO_2 and O_2. The charge transfer and gas-induced charge fluctuation significantly affected the transport properties of SWCNTs, which are consistent with previous experimental results[102–105].

On the experimental side, Lau et al.[106] functionalized MWCNTs using a range of oxidative techniques, including thermal treatment, acid reflux, and dry UV/O_3 treatment. The effects of these treatments on the structure and electrical properties were characterized. The results confirmed that functionalization enhanced the electrical conductivity of MWCNTs, and the electron transfer from the carbon atoms on MWCNTs to functional groups attached to the surface was mainly responsible for the increased conductivity. The traditional acid reflux technique was also shown to increase the conductivity MWCNTs, but the primary structure of MWCNTs was greatly modified by the introduction of kinks and bends, which probably limited the increase in conductivity. Thermal treatment was also found to be effective in introducing new functional groups on CNTs, and in most cases, this process was less harmful in reducing the nanotube length and introducing defects than acid treatment. A major conclusion of this study was that dry UV/O_3 was the best among the three techniques in terms of enhancing electrical conductivity, even though the degree of functionalization was low, likely due to less damage to nanotubes on different length scales, which in turn encouraged the formation of networks between the nested tube structures.

The effects of amino functionalization using a ball milling technique on the electrical and electronic properties of CNTs were evaluated[75,76]. It was noted that with an increase in the ball milling time, the electrical conductivity of CNTs milled in the presence of NH_4HCO_3 increased gradually—a remarkable 250% increase after 9 hours of milling was achieved. In sharp contrast, no obvious increase or even a reduction in electrical conductivity was found in the CNTs milled without this chemical (Figure 3.13A). The temperature-dependent conductance of CNTs after ball milling with and without NH_4HCO_3 showed similar trends (Figure 3.13B). For the as-received CNTs and CNTs subjected to ball milling only, the conductance increased continuously as temperature increased (curves a and b in Figure 3.13B), representing electrical behavior of a typical semiconductor. For the CNTs milled in the presence of NH_4HCO_3, the increase in conductivity was more pronounced (curve c in Figure 3.13B) than the other CNTs, suggesting a higher conductivity for a given temperature. These contrasting behaviors strongly indicated that NH_4HCO_3 played an important role in the improvement of electrical conductivity of CNTs after ball milling. Ball milling conducted in the absence of chemicals created defects and amorphous materials[75,76,107–110], resulting in a negative effect on electrical properties of CNTs due to the destruction of graphitic layers and perturbation of the π electron system on the CNT walls. To explain the beneficial effect of NH_4HCO_3, the carrier concentrations of CNTs ball milled with and without this chemical were measured as summarized in Table 3.4.

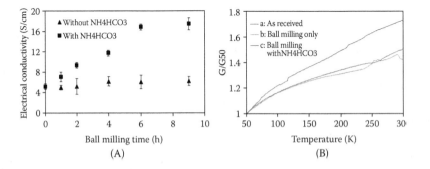

FIGURE 3.13
Electrical conducting behavior of CNTs: (A) electrical conductivity of CNTs as a function of ball milling time, and (B) temperature-dependent conductance of CNTs after ball milling with and without NH_4HCO_3 for 2 hours. Adapted from Ma, P. C., et al. 2008. *Chem. Phys. Lett.* 458: 166.

TABLE 3.4 Carrier Type and Concentration of CNTs after Amino Functionalization.

Sample	Without NH_4HCO_3 (cm^{-3})	With NH_4HCO_3 (cm^{-3})
Pristine CNTs	$+ (3.28\pm0.27)\times10^{20}$	$+ (3.28\pm0.27)\times10^{20}$
1 h	$+ (1.93\pm1.55)\times10^{20}$	$- (1.74\pm0.77)\times10^{19}$
2 h	$+ (1.98\pm0.78)\times10^{20}$	$- (6.84\pm1.02)\times10^{19}$
4 h	$+ (2.75\pm0.31)\times10^{20}$	$- (4.17\pm1.47)\times10^{20}$
6 h	$+ (2.62\pm1.20)\times10^{20}$	$- (6.54\pm0.88)\times10^{20}$
9 h	$+ (1.63\pm1.21)\times10^{20}$	$- (9.39\pm1.25)\times10^{20}$

Source: Adapted from Ma, P. C., et al. 2008. *Chem. Phys. Lett.* 458: 166.

The pristine CNTs were a kind of semiconductor with a positive carrier concentration. A marginal reduction in carrier concentration was noted for the CNTs milled in the absence of NH_4HCO_3, suggesting some damage to the CNT structure. The electrical carrier of these CNTs was positive, being identical to the pristine CNTs. It is remarkable to note that the carrier concentration became negative upon ball milling in the presence of NH_4HCO_3, and its absolute value gradually enhanced with increased milling time. The positive and negative carrier concentrations were closely related to different types of semiconductors, that is, p-type and n-type, respectively: the pristine CNTs and those obtained after ball milling without NH_4HCO_3 were p-type semiconductors, whereas those produced after ball milling with NH_4HCO_3 were n-type. It is well known that the major carriers are holes with positive charges and electrons with negative charges, respectively, for the p-type and n-type semiconductors. The latter CNTs with negative charges contained amine and amide functional groups attached on the surface (Figure 3.7). Because the nitrogen atoms in these functional groups contained lone-pair electrons that functioned as the electron donor, charge transfer occurred between these groups and the CNTs[111], resulting in the conversion of the semiconducting behavior of the CNTs from p-type to n-type along with a simultaneous increase in electrical conductivity (Figure 3.13).

Converting the semiconducting behavior of CNTs can find many potential applications, for example, rectifying p–n junctions, bipolar junctions, photovoltaic devices, and field-effect transistors, all of which require accurate control of electrical properties for different functions. Apart from being able to convert the semiconducting behavior of CNTs, the above in-situ functionalization technique also holds unique advantages, such as simplicity and ease of application as well as cost effectiveness.

3.6.4 Thermal Properties

Thermal conductivity and thermal stability are the two most important parameters that describe the thermal properties of CNTs. The thermal conductivity of CNTs is governed by the transportation of phonons (see

Section 1.2), thus, any functionalization processes having adverse effects on phonon transportation will decrease the thermal conductivity of CNTs.

Pan et al.[112] used a nonequilibrium molecular dynamics method to study the thermal conductivities of hydrogenated (10, 0) SWCNTs. The simulations revealed that the thermal conductivity of pristine (10, 0) CNTs was about 237 W/(m − K), which is comparable to the experiment value. When 5% carbon atoms were hydrogenated, the thermal conductivity of CNTs was reduced by a factor of 1.5. Additional functionalization with hydrogen atoms further reduced their thermal conductivity (Figure 3.14). The degradation of thermal conductivity of functionalized CNTs arose from the reduction of the phonon scattering length and the suppression of some thermal vibration modes due to chemical attachment of hydrogen atoms. Indeed, the phonon spectra of the functionalized CNTs revealed a significant change in the density of phonon modes. Furthermore, the thermal conductivities of SWCNTs were much less influenced by physical or noncovalent functionalization than chemical or covalent functionalization.

However, very little information is available so far about the experimental measurement of thermal conductivities of functionalized CNTs, although significant progress has been made on the same topic for CNT/polymer nanocomposites.

CNTs also exhibit excellent thermal stability, and there is no significant weight loss for purified CNTs above 550°C, even in air flow (see Figure 1.8). Functionalized CNTs in general show lower thermal degradation temperatures, as most functionalization processes involve either damage to CNT structure or introduction of organic materials on CNTs. The damaged CNTs

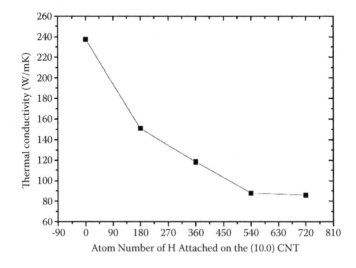

FIGURE 3.14
Thermal conductivity versus atom number of hydrogen atoms attached to (10, 0) CNTs at 300 K, which is determined by a nonequilibrium molecular dynamics simulation. Adapted from Pan, R., et al. 2007. *Nanotechnology* 18: 285704.

result in the production of amorphous carbon particles that can be burned easily in air flow, and the functional moieties attached on CNT covalently or noncovalently are also easily burned. One advantage of carrying out the TGA analysis is that one can evaluate the weight percentage of heterogeneous atoms on CNTs, thus offering an indirect way to measure the degree of CNT functionalization for controlling the functionalities, structure, and properties of CNTs.

3.7 Nanoparticle (NP)/CNT Nanohybrids

3.7.1 Introduction to NPs

Nanoparticles (NPs) have been used as filler materials in the plastics industry. NPs are basically additives in solid form, which differ from the polymer matrix in terms of their composition and structure. They generally comprise inorganic materials, such as metal and metal oxides, ceramic compounds, carbon-based particles, and so on.

Metal NPs are particularly interesting nanoscale systems because of the ease with which they can be synthesized and modified chemically. Compared to bulk metals, metal NPs exhibit a number of unique properties, including a large surface-to-volume ratio, high surface reaction activity, high catalytic efficiency, and strong adsorption ability[113]. For example, when bending of bulk copper occurs, copper atoms/clusters move at about the 50-nm scale. Copper NPs smaller than 50 nm are considered super-hard materials that do not exhibit the same malleability and ductility as bulk copper. Suspensions of NPs are possible because the interaction of the particle surface with the solvent is strong enough to overcome differences in density, which usually result in either sinking or floating of the material in a liquid. Metal NPs often have unexpected visible properties because they are small enough to confine their electrons and produce quantum effects. For example, gold NPs appear deep red to black in a solution. Due to these unique properties, metal NPs have been widely used as catalysts, conducting fillers, and electronic and optical materials[114].

Preparation of metal NPs can be classified broadly into two categories: (1) subdivision of bulk metal materials using physical or mechanical methods; and (2) growth of particles from metal atoms generated from corresponding metal precursors using chemical reactions (Figure 3.15A). The latter chemical procedure becomes more and more predominant due to the ease of preparation and the possibility of mass production. It does not require special and expensive apparatuses such as vacuum systems, and is very suitable for small and uniformly sized metal NPs. The polyol process is a typical example of the chemical procedure, where metallic salts are dissolved in a liquid polyol (mostly ethylene glycol) and controlled heating is applied.

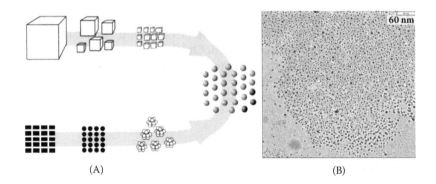

(A)   (B)

FIGURE 3.15
(A) Schematic illustration of the preparation of metallic NPs, and (B) TEM image of Ag NPs fabricated by a chemical method. Adapted from Kim, J. K., et al. 2008. U.S. Provisional Patent Application No. 61/131,222.

At an elevated temperature, polyol is decomposed to aldehyde, resulting in the reduction of metal ions to form metal particles. The processing conditions have a large impact on the size distribution, degree of agglomeration, and crystallinity of metal NPs[117,118]. Different metal NPs, such as Ag, Au, Pd, Pt, Ru, Rh, Ir, Os, Ni, Fe, Co, Ni, Ag-Pd alloy, Pd-Cu alloy, and so on, have been successfully developed by employing this method[117–120]. Figure 3.15B shows the TEM image of Ag NPs fabricated by a polyol process[116]. The size distribution of these NPs is quite uniform, and the diameter of most NPs is less than 10 nm.

A ceramic material may be defined as an inorganic and nonmetallic solid material. Bulk ceramic materials are brittle, hard, and strong in compression, but weak in shear and tension. They exhibit an excellent ability to withstand chemical erosion and high temperatures. However, when the size of ceramic particles is reduced from microns (> 10 μm) to 10–100 nm, the percentage of atoms locate in the intercrystalline phase, that is, grain boundaries and triple junctions become increasingly higher. Thus, the materials are in a nonequilibrium state, exhibiting chemical, optical, thermal, magnetic, electrical, and mechanical properties that are totally different from their bulky state[121]. For example, oxide ceramic NPs have acidic or basic properties due to the fact that their surface contains hydroxyl (–OH) groups, and these groups play an important role in anchoring some medical and biological molecules, making them useful for application in medicine and biotechnology.

Many methods are available for the preparation of ceramic NPs. These can be classified into three basic types: mechanical, chemical, and vapor–phase methods[121–123]. The mechanical method is suitable for producing NPs from coarse grains that are derived from naturally occurring minerals. In this method, milling is applied until the desired particle size is obtained. The limitation of this method is that the minimum particle size can only be at around 100 nm, although smaller particles with diameters of around 50 nm can be achieved by employing high-energy ball milling. In addition, some impurities can be introduced into the NPs, mainly from the milling beads and containers. Therefore, this method is popularly used to produce traditional ceramic NPs where high purity and small diameters of NPs are not required and cost is one of the most important requirements. The chemical method, such as sol-gel processing, offers several advantages over the mechanical method because the chemical method can accurately control the particle morphology and purity. Synthesized NPs may require subsequent thermal treatments for dehydration, removal of organics, and controlled crystallization to form oxides with desirable structure and crystallite size. The chemical process is widely used in the production of advanced ceramic NPs, such as aluminum oxide, silicon carbide, and silicon nitride, with diameters up to several nanometers.

The vapor–phase techniques are relatively expensive for producing NPs, but offer many advantages over other methods, such as the ability to produce NPs with high purity, narrow size distribution, and nonaggregation. In these techniques, the reaction mixture is heated above the boiling point of a solvent in a closed system and condensed in the gas phases. The condensate is transported by convection to the liquid nitrogen cold fringe. The clusters are scraped from the cold fringe and collected via a funnel. It is possible to have the particles transferred directly into a cold press where they can be compacted. With this technique, TiO_2, TiN, AlN, TiC, Ti/C/N, and Al_2O_3 NPs with diameters as small as 10–20 nm can be produced.

3.7.2 Fabrication of NP/CNT Nanohybrids

Both CNTs and NPs have emerged as new classes of materials that are particularly of interest to all material scientists and engineers. Furthermore, the combination of these two classes of materials may create completely new hybrid materials whose properties are unique and different from those of the individual components acting alone. In these nanohybrid materials, the CNT surface serves as a template onto which NPs are absorbed or, when bearing functional groups, CNTs may be linked through organic fragments to metallic or semiconducting NPs[124].

In a report on decorating CNTs with metallic NPs, Planeix et al.[125] described the use of SWCNTs as a supporting material for the decoration of Ru NPs that can act as catalysts in heterogeneous catalysis. Ru 2, 5-pentanedionate was spread onto the walls of SWCNTs, which was subsequently reduced under a stream of diluted hydrogen (H_2:N_2=1:9 molar ratio). The Ru NPs thus obtained were well dispersed on the nanotube surface as corroborated by a detailed TEM analysis. The weight content of Ru NPs in the final material was about 0.2%. Catalytic assays including liquid-phase hydrogenation of cinnamaldehyde revealed a particularly high selectivity for cinnamyl alcohol (up to 92%) with an 80% conversion of cinnamaldehyde. In contrast, under similar conditions, Ru NPs of similar size supported on Al_2O_3 catalyze the formation of cinnamyl alcohol with a selectivity of 20–30% only, suggesting a unique advantage of NP/CNT nanohybrids for catalytic reactions.

Following the previously mentioned promising work, many CNT-based nanohybrids have been developed with either metallic or semiconducting NPs. There are two main pathways to obtain nanohybrids: (1) NPs are grown or deposited directly onto the CNT surface and (2) NPs are fabricated at first, followed by connection to CNTs using linkers through chemical or physical interactions.

For the direct formation of NPs on the CNT surface, salts of metals are commonly used as precursors for NPs, which are obtained by a reduction process. If this process occurs in the presence of CNTs, the NPs can be deposited onto the CNT walls, mostly through van der Waals interactions that are not strong enough to guarantee meaningful adhesion. Various methods, such as the thermal decomposition reaction[126], vapor deposition[127], surface chemical reduction[128], and gamma irradiation[129], have been utilized to perform the metal cation reduction. Precious and noble metals, such as Pt, Au, Pd, Ag, Rh, and Ru, as well as transitional metals, such as Ni, Co, and Fe, have been extensively studied in the literature[124]. These metals are commonly used in heterogeneous catalytic reactions and their properties can be enhanced when CNTs are employed as support material.

Xue and coworkers[126] also proposed methods to grow Pt, Ag, Au, Pd, and Cu NPs on MWCNTs. The metal cations were initially dispersed onto the nanotube surface and then reduced under a hydrogen stream while heating. Besides acting as supporting material, CNTs act as a template in tailoring the size of metal particles. This hypothesis was further supported by the fact that the NPs were much larger when the reduction of the metal cations was performed in the presence of graphite or amorphous carbon materials. Moreover, the number of NPs deposited on graphite or active carbon using the same method was found to be very small compared to that obtained in the presence of CNTs.

However, there are still two major drawbacks identified in applying these methods to fabricate well-defined NP/CNT nanohybrids—weak interactions between the NPs and the CNT surface[130,131], and agglomeration of NPs[124]. In most cases, NPs are formed inside the tubes due to the capillary effect[132,133], leading to a poor synergic effect of the NP/CNT nanohybrids. The agglomeration of NPs causes nonuniform decoration of CNTs. To realize the effective production of NP/CNT nanohybrids, CNTs need to be functionalized to enhance their interactions with NPs[124]. Ma et al.[75,76] used a ball milling technique to introduce amino functional groups on CNT surfaces, and studied the interactions between Ag-NPs and functionalized CNTs[134]. It was noted that when pristine CNTs were used as a template for Ag decoration, most Ag-NPs were deposited onto the carbon film of the copper grid used for TEM characterization (Figure 3.16A). High-resolution TEM (HRTEM) images confirmed that the CNT surface was relatively clean without any evidence of Ag-NPs on it. Several Ag-NPs were found inside the CNTs, which were introduced due to the capillary effect (Figure 3.16B).

In sharp contrast, when the amino-functionalized CNTs were employed for Ag decoration, very different results were obtained: Figures 3.16C and D present a number of tiny Ag-NPs attached to the CNT surface. The Ag-NPs were uniformly distributed without agglomeration, and their sizes were also uniform, ranging from subnanometer to about 2–4 nm. The strong signal at about 3 keV of the energy dispersive x-ray (EDX) spectrometer spectrum (Figure 3.16E) obtained from the TEM was a testament to the decoration of Ag-NPs on the CNT surface. The mechanism behind the successful Ag-NP decoration on the CNT surface after functionalization is schematically illustrated in Figure 3.17. During the functionalization of CNTs using ball milling, ammonia gas was introduced into the system via decomposition of NH_4HCO_3. The ball milling also introduced defects and broke the –C–C– bonds on the surface layer of the CNTs, which was further facilitated by the presence of NH_4HCO_3[75,76]. These CNT surface modifications, in turn, allowed the amine groups to form covalent bonds with the broken –C–C– bonds. The coordination interaction between the Ag cations and amine groups resulted in the attachment of Ag particles on the functionalized CNTs, as illustrated in Figure 3.16.

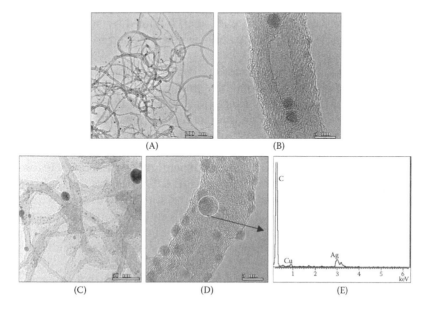

(A) (B)

(C) (D) (E)

FIGURE 3.16
TEM images of the Ag@CNTs (A) without and (C) with functionalization and the corresponding HRTEM (B and D) and (E) EDX spectrum (spot A in D) Adapted from Ma, P. C., et al. 2008. *Carbon* 46: 1497.

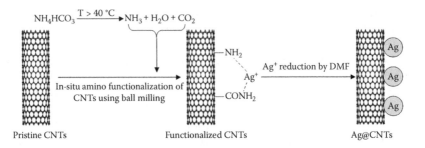

FIGURE 3.17
Schematics of CNT functionalization and interaction with Ag cations. Adapted from Ma, P. C., et al. 2008. *Carbon* 46: 1497.

Besides direct deposition of NPs on CNTs, they can be linked to the CNT surface via covalent bonds and intermolecular interactions such as π-π stacking, and hydrophobic or electrostatic attractions[124].

Covalent linkage of NPs with CNTs is the most commonly used technique to fabricate NP/CNT nanohybrids. In this method, CNTs are first functionalized to endow some reactive end groups, such as thiol, amine, and carboxyl. Prepared NPs are then introduced into the CNT solution to react with these groups via the formation of covalent bonds. The so-called Bingel reaction, that is, the formation of Au-S covalent bonds, is a typical example (Reaction A in Figure 3.18) of this kind of reaction. This method was first developed[22] where CNTs were functionalized with acid, followed by subsequent reactions with aliphatic amines. The Au-NPs were connected to the functionalized CNTs.

Hydrophobic and hydrogen interactions between the ligands have also been used to immobilize the metal NPs onto CNTs where monolayers are formed to passivate the metal surface (Reaction B in Figure 3.18). Acetone molecules adsorbed on CNTs can interact hydrophobically with Au nanoclusters covered with a monolayer of octanethiols[135]. Au-NPs with a size of 2 nm or 5 nm and covered by a mixed-monolayer of decanethiol and mercaptoundecanoic acid, were linked efficiently to oxidized CNTs due to the presence of carboxylic groups on the surface[136]. Besides the hydrophobic interactions between the alkyl chains and the CNT surface, hydrogen bonds between the carboxylic groups of CNTs and those present on the NP surface resulted in very stable composite materials. These systems can be potentially important for creating a rich variety of molecular nanostructures for device applications.

The graphene layer of CNTs is a kind of aromatic ring that can interact physically with the compounds containing aromatic rings via π-π stacking. Among the aromatic compounds, molecules bearing a long aliphatic chain terminated by a thiol group are widely employed to stack onto the CNT surface[137,138]. The end groups of pyrene derivatives can react with some metal NPs, as shown in Figure 3.18 (Reaction C). Spectroscopic experiments on the

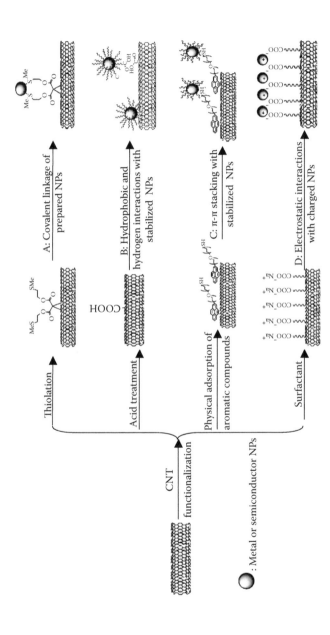

FIGURE 3.18
Techniques to connect NPs with the CNT surface. Adapted from Han, L. et al. 2004. Langmuir 20: 6019.

TABLE 3.5 Summary of the Major Applications of NP/CNT Nanohybrids.

Application	NPs	Advantages
Catalyst	Co, Pd, Pt, Au, Ag, Metal alloy (Ru/Sn, Ru/Pt)	High activity and excellent regioselectivity in some chemical reactions.
Hydrogen storage and sensing	Ni, Pt	Reversible chemisorption of hydrogen than CNT alone, better hydrogen storage and sensing capability (limited detection of around 400 ppm).
Optical and electronic applications	TiO_2, CdSe, CdTe, CdSe, ZnS, MoO_2 Ni, Fe, Co, Au, RuO_2/$W(CO)_6$	Higher conversion rates on electronic structure, carrier trapping, and delocalization of CNT-based photovoltaic devices. Effective visualization of the size and manipulation of clusters of CNTs using optical detection. Formation of controlled electronic contacts between conducting CNTs. Field emitters: lower threshold voltages and higher amplification factors compared to CNT field emitters alone; possible applications in field-emitter displays.
Gas sensor and biosensor	Pd Pt	Increased sensitivity and power consumption with a lower detection limit over a good linear range. Limiting the interfacial effect on detection compared to CNT alone.
Fillers for electrically conducting polymer nanocomposites	Ag	Enhanced electrical/thermal conductivities of CNT/polymer nanocomposites with balanced mechanical properties; reduced contact resistance of CNT networks formed inside the polymer matrix.
Energy storage	Co, Pt and Ru	Enhanced electrochemical reduction in aqueous solution, along with higher power efficiency in fuel cells. Enhanced electrical/thermal conductivity and lower thermal expansion used as electrode materials in lithium-ion batteries.

Sources: Ma, P. C., et al. 2008. *Carbon* 46: 1497; Ellis, A. V., et al. 2003. *Nano. Lett.* 3: 279; Han, L., et al. 2004. *Langmuir,* 20, 6019; Liu, L. Q., et al. 2003. *Chem. Phys. Lett.* 367: 747; Georgakilas, V., et al. 2005. *Chem. Mater.* 17: 1613; Stoffelbach, F., et al. 2005. Chem. Commun. 36: 4532; and Wildgoose, G. G., et al. 2005. *Small* 2: 182.

composite material indicated that the pyrene fluorescence was completely quenched and the Raman spectrum of the CNT was enhanced, a reflection of charge transfer between the CNTs and the metal Au-NPs mediated by the linker. Electron transfer processes between the donor groups and CNTs are also known to occur on CNT-Zn porphyrin (containing aromatic rings) hybrid materials.

Electrostatic interactions have also been used to anchor metal colloids to CNTs. CNTs are treated with a surfactant or an ionic polyelectrolyte which serves as an anchor for charged metal NPs (Reaction D in Figure 3.18). By choosing different kinds of modifiers, the surface of CNTs can be negatively charged in order to accept positively charged NPs. For example, Stoffelbach et al.[139] devised an easy route to decorate MWCNTs with magnetic NPs, like Fe_3O_4. The two components form a stable composite by electrostatic interaction. The CNTs were functionalized with a surfactant (poly(2-vinylpyridine), resulting in a larger increase of grafting density of carboxylate groups onto the CNT surface than oxidative methods, which also caused a severe reduction in CNT length. The magnetite NPs were treated with a HNO_3 solution to obtain positive charges on their surface. The functionalized CNTs and charged NPs were dispersible in water, resulting in the decoration of Fe_3O_4 NPs on the MWCNTs.

3.7.3 Applications of NP/CNT Nanohybrids

A myriad of studies involving the fabrication and application of NP/CNT nanohybrids have been reported in recent years. The applications of these nanohybrids are numerous, ranging from catalysts to supercapacitors, and a summary is presented in Table 3.5. In-depth discussion on the applications of these nanohybrids can be found elsewhere[134–140].

References

1 Hirsch, A., et al. 2005. *Top Curr. Chem.* 245: 193.
2 Basiuk, E. V., et al. 2004. *Nano. Lett.* 4: 863.
3 Balasubramanian, K., et al. 2005. *Small* 1: 180.
4 Mylvaganam, K., et al. 2004. *J. Phys. Chem. B* 108: 5217.
5 Chen, Z. F., et al. 2003. *Chem. Phys. Chem.* 4: 93.
6 Nakajima, T., et al. 1996. *Eur. J. Solid State Inorg. Chem.* 33: 831.
7 Hamwi, A., et al. 1997. *Carbon* 35: 723.
8 Mickelson, E. T., et al. 1998. *Chem. Phys. Lett.* 296: 188.
9 Mickelson, E. T., et al. 1999. *J. Phys. Chem. B* 103: 4318.
10 Marcoux, P. R., et al. 2002. *Phys. Chem. Chem. Phys.* 4: 2278.
11 Touhara, H., et al. 2002. *J. Fluor. Chem.* 114: 181.
12 Gu, Z., et al. 2002. *Nano. Lett.* 2: 1009.
13 Boul, P. J., et al. 1999. *Chem. Phys. Lett.* 310: 367.
14 Saini, R. K., et al. 2003. *J. Am. Chem. Soc.* 125: 3617.
15 Stevens, J. L., et al. 2003. *Nano. Lett.* 3: 331.
16 Unger, E., et al. 2002. *Curr. Appl. Phys.* 2: 107.
17 Hou, P. X., et al. 2002. *Carbon* 40: 81.
18 Chen, Y., et al. 1998. *J. Mater. Res.* 13: 2423.
19 Pekker, S., et al. 2001. *J. Phys. Chem. B* 105: 7938.

20　Owens, F. J., et al. 2002 (December). 23rd Army Science Conference, Orlando, FL, Session L/LP-11.
21　Yildirim, T., et al. 2001. *Phys. Rev. B* 64: 075404.
22　Chen, J., et al. 1998. *Science* 282: 95.
23　Hirsch, A. 2002. *Angew. Chem. Int. Ed.* 41: 1853.
24　Tasis, D., et al. 2006. *Chem. Rev.* 106: 1105.
25　Peng, X., et al. 2009. *Adv. Mater.* 21: 625.
26　Ni, B., et al. 2000. *Phys. Rev. B* 61: R16343.
27　Bahr, J. L., et al. 2001. *Chem. Mater.* 13: 3823.
28　Mitchell, C. A., et al. *Macromolecules* 35: 8825.
29　Hudson, J. L., et al. 2004. *J. Am. Chem. Soc.* 126: 11158.
30　Dyke, C. A., et al. 2003. *J. Am. Chem. Soc.* 125: 1156.
31　Peng, H., et al. 2003. *Chem. Commun.* 3: 362.
32　Umek, P., et al. 2003. *Chem. Mater.* 15: 4751.
33　Ying, Y., et al. 2003. *Org. Lett.* 5: 1471.
34　Liang, F., et al. 2004. *Nano. Lett.* 4: 1257.
35　Penicaud, A., et al. 2005. *J. Am. Chem. Soc.* 127: 8.
36　Borondics, F., et al. 2005. *Fullerenes Nanotubes Carbon Nanostruct.* 13: 375.
37　Khare, B. N., et al. 2004. *J. Phys. Chem. B* 108: 8166.
38　Liu, P. 2005. *Eur. Polym. J.* 41: 2693.
39　Ma, P. C. 2008. Novel surface treatment, functionalization and hybridization of carbon nanotubes and their polymer-based composites. PhD dissertation. Hong Kong University of Science and Technology. 1–15.
40　Vaisman, L., et al. 2006. *Adv. Colloid Interface Sci.* 128–130: 37.
41　Grossiord, N., et al. 2006. *Chem. Mater.* 18: 1089.
42　McCarthy, B., et al. 2001. *Synth. Met.* 121: 1225.
43　Hill, D. E., et al. 2002. *Macromolecules* 35: 9466.
44　Panhuis, M., et al. 2003. *J. Phys. Chem. B* 107: 478.
45　O'Connell, M. J., et al. 2001. *Chem. Phys. Lett.* 342: 265.
46　Della, N. F., et al. 2003. *Fullerenes Nanotubes Carbon Nanostruct.* 11: 25.
47　Kang, Y. K., et al. 2009. *Nano. Lett.* 9: 1414.
48　Curran, S. A., et al. 1998. *Adv. Mater.* 10: 1091.
49　Coleman, J. N., et al. 2000. *Adv. Mater.* 12: 213.
50　Yi, W., et al. 2008. *J. Phys. Chem. B* 112: 12263.
51　Zheng, M., et al. 2003. *Nature Mater.* 2: 338.
52　Lin, Y., et al. 2004. *J. Mater. Chem.* 14: 527.
53　Daniel, S., et al. 2007. *Sens. Actuators B* 122: 672.
54　Lu, F., et al. 2009. *Adv. Mater.* 21: 139.
55　Jiang, K., et al. 2004. *J. Mater. Chem.* 14: 37.
56　Williams, K. A., et al. 2002. *Nature* 420: 761.
57　Heller, D. A., et al. 2006. *Science* 311: 508.
58　Lu, Y., et al. 2006. *J. Am. Chem. Soc.* 128: 3518.
59　Liu, Z., et al. 2007. *Angew Chem. Int. Ed.* 46: 2023.
60　Rege, K., et al. 2006. *Small* 2: 718.
61　Zheng, M., et al. 2003. *Science* 28: 1545.
62　Chen, R. J., et al. 2001. *J. Am. Chem. Soc.* 123: 3838.
63　Nepal, D., et al. 2007. *Small* 3: 1259.
64　Wei. G., et al. 2010. *Carbon* 48: 645.
65　Poenitzsch, V. Z., et al. 2007. *J. Am. Chem. Soc.* 129: 14714.

66 Lacerda, L., et al. 2007. *Nano Today* 2: 38.
67 Pantarotto, D., et al. 2004. *Angew Chem. Int. Ed*. 43: 5242.
68 Lopez, C. F., et al. 2004. *Proceedings of the National Academy of Sciences* (*PNAS*) 101: 4431.
69 Kateb, B., et al. 2007. *NeuroImage* 37: S9.
70 Rojas-Chapana, J., et al. 2005. *Lab Chip* 5: 536.
71 Bandyopadhyaya, R., et al. 2002. *Nano. Lett.* 2: 25.
72 Star, A., et al. 2002. *Angew Chem. Int. Ed*. 14: 2508.
73 Kim, O. K., et al. 2003. *J. Am. Chem. Soc.* 125: 4426.
74 Wang, H., et al. 2006. *J. Am. Chem. Soc.* 128: 13364.
75 Ma, P. C., et al. 2008. *Chem. Phys. Lett.* 458: 166.
76 Ma, P. C., et al. 2009. *J. Nanosci. Nanotechnol.* 9: 749.
77 Kónya, Z., et al. 2002. *Chem. Phys. Lett.* 360: 429.
78 Pan, H. L., et al. 2003. *Nano. Lett.* 3: 29.
79 Li, X., et al. 2003. *Chem. Phys. Lett.* 377: 32.
80 Sham, M. L., et al. 2009. *J. Compos. Mater.* 43: 1537.
81 The Ozone Hole, http://www.theozonehole.com/ozonecreation.htm (accessed December 2009).
82 Poulis, J. A., et al. 1993. *Int. J. Adhes. Adhes.* 13: 89.
83 Macmanus, L. F., et al. 1999. *J. Polymer Sci. A* 27: 2489.
84 Sham, M. L., et al. 2006. *Carbon* 44: 768.
85 Ma, P. C., et al. 2007. *Compos. Sci. Technol.* 67: 2965.
86 Li, J., et al. 2005. *Scripta Mater.* 53: 235.
87 Li, J., et al. 2007. *Compos. Sci. Technol.* 67: 296.
88 Ago, H. 1999. *J. Phys. Chem. B* 103: 8116.
89 Chen, C., et al. 2010. *Carbon* 48: 939.
90 Ma, P. C., et al. 2006. *Carbon* 44: 3232.
91 Ma, P. C., et al. 2010. *Carbon* 48: 1824.
92 Sham, M. L., et al. 2004. *IEEE Trans. Adv. Packaging* 27: 179.
93 Hoecker, F., et al. 1998. *J. Appl. Polym. Sci.* 59: 139.
94 Neimark, A. V. J. 1999. *Adhes. Sci. Technol.* 13: 1137.
95 Nuriel, S., et al. 2005. *Chem. Phys. Lett.* 404: 263.
96 Weldon, D. G. 2009. *Failure analysis of paints and coatings* (Rev. ed), 13. New York: John Wiley & Sons.
97 Garg, A., et al. 1998. *Chem. Phys. Lett.* 295: 273.
98 Zhang, Z. Q., et al. 2008. *Nanotechnology* 19: 395702.
99 He, X. Q., et al. 2009. *J. Phys. Condens. Matter* 21: 215301.
100 Park, H., et al. 2006. *Nano. Lett.* 6: 916.
101 Zhao, J., et al. 2002. *Nanotechnology* 13: 195.
102 Kong, J., et al. 2000. *Science* 287: 622.
103 Collins, P. G, et al. 2000. *Science* 287: 1801.
104 Tang, X. P., et al. 2000. *Science* 288: 492.
105 Sumanasekera, G. U., et al. 2000. *Phys. Rev. Lett.* 85: 1096.
106 Lau, C. H., et al. 2008. *J. Nanopart. Res.* 10: 77.
107 Huang, J. Y., et al. 1999. *Chem. Phys. Lett.* 303: 130.
108 Kim, Y. A., et al. 2002. *Chem. Phys. Lett.* 355: 279.
109 Liu, F., et al. 2003. *Carbon* 41: 2527.
110 Li, X., et al. 2007. *Chem. Phys. Lett.* 444: 258.
111 Klinke, C., et al. 2005. *Nano. Lett.* 5: 555.

112 Pan, R., et al. 2007. *Nanotechnology* 18: 285704.
113 Nanoparticle. *Wikipedia,* http://en.wikipedia.org/wiki/Nanoparticle (accessed January 2010).
114 Feldheim, D. L., et al. 2001. *Metal NPs: Synthesis, characterization, and applications,* 261. Boca Raton, FL: CRC Press.
115 Kumar, C. 2007. *Nanomaterials for biosensors,* 279. New York: Wiley-VCH.
116 Kim, J. K., et al. 2008. U.S. Provisional Patent Application No. 61/131,222.
117 Figlarz, M., et al. 1985. U.S. Patent No. 4539041.
118 Goia, D. V., et al. 2006. U.S. Patent No. 20060090599.
119 Waseda, Y., et al. 2004. *Morphology control of materials and NPs: Advanced materials processing and characterization,* 85. New York: Springer.
120 Sugimoto, T. 2000. *Fine particles: Synthesis, characterization, and mechanisms of growth,* 460. Boca Raton, FL: CRC Press.
121 Schubert, U., et al. 2005. *Synthesis of inorganic materials,* 355. New York: Wiley-VCH.
122 Carter, C. B., et al. 2007. *Ceramic materials: Science and engineering,* 365. New York: Springer.
123 Koch, C. C. 2007. *Nanostructured materials: Processing, properties, and applications,* 16. Norwich, NY: William Andrew Publishing.
124 Han, L. et al. 2004 Langmuir 20: 6019.
125 Planeix, J. M., et al. 1994. *J. Am. Chem. Soc.* 116: 7935.
126 Xue, B., et al. 2001. *J. Mater. Chem.* 11: 2378.
127 Guo, D. J., et al. 2005. *Carbon* 43: 1259.
128 Liu, Y., et al. 2006. *Carbon* 44: 381.
129 Oh, S. D., et al. 2005. *Mater. Lett.* 59: 1121.
130 Durgun, E., et al. 2003. *Phys. Rev. B* 67: 201401.
131 Zamudio, A., et al. 2006. *Small* 2: 346.
132 Ajayan, P. M., et al. 1996. *Nature* 361: 333.
133 Ugarte, U., et al. 1996. *Science* 274: 1897.
134 Ma, P. C., et al. 2008. *Carbon* 46: 1497.
135 Ellis, A. V., et al. 2003. *Nano. Lett.* 3: 279.
136 Han, L., et al. 2004. *Langmuir,* 20: 6019.
137 Liu, L. Q., et al. 2003. *Chem. Phys. Lett.* 367: 747.
138 Georgakilas, V., et al. 2005. *Chem. Mater.* 17: 1613.
139 Stoffelbach, F., et al. 2005. Chem. Commun. 36: 4532.
140 Wildgoose, G. G., et al. 2005. *Small* 2: 182.

4

CNT/Polymer Nanocomposites

4.1 Introduction

4.1.1 Polymer Nanocomposites

Polymer composites, consisting of additives and polymer matrices including thermoplastics, thermosets, and elastomers, are considered to be an important group of relatively inexpensive materials for many useful applications. Constituent materials with different properties are selected to fabricate composites to improve one or more properties: for example, high-modulus carbon fibers are added to polymer matrices to fabricate composites that have enhanced mechanical and fracture properties. However, there are bottlenecks in optimizing the properties of polymer composites by employing traditional micrometer scale fillers. The conventional filler content in polymer composites is generally in the range of 10 to 70 wt%, giving rise to a higher density than that of the neat polymer matrix. In addition, stiffness of composites is often traded for toughness, and microscopic defects and voids arising from the high volume fraction of filler often lead to premature failure of the composites[1].

In contrast to the traditional polymer composites containing microscale fillers, the incorporation of nanoscale carbon nanotubes (CNTs) as an additive in a polymer resin results in CNT/polymer nanocomposites. The sheer size of CNTs means a very small particle distance, which in turn influences the properties of nanocomposites even at an extremely low CNT content. For example, the electrical conductivity of CNT/epoxy nanocomposites increased by several orders of magnitude with less than 0.5 wt% of CNTs[2]. The excellent physical and mechanical properties combined with many desirable functional properties of CNTs are finding huge potential applications of CNT/polymer nanocomposites. In addition, CNT/polymer nanocomposites are one of the most extensively studied composite systems partly because low-cost processes are widely available for manufacturing of the composites.

In CNT/polymer nanocomposites, dispersion of CNTs is one of most critical and important issues because the degree of dispersion determines the various properties of the resulting composites. As discussed in Chapter 2, the uniform dispersion of nanoscale particles in a viscous polymer resin is extremely difficult relative to the separation of microscale particles. Characterizing the degree of dispersion is also difficult because of the small size of nanoparticles. The final dispersion of CNTs in a polymer is not only a function of the method used to separate individual tubes, but also a function of the method used to mix the nanotubes with polymers and the ensuing curing and forming processes involved in the final nanocomposites.

4.1.2 Classification of CNT/Polymer Nanocomposites

Following the first report on the preparation of CNT/polymer nanocomposites in 1994[3], a number of research papers have appeared in open media to study their interesting properties, including the structure–property relationship and potential applications in many different fields. The number of publications directly relevant to CNT/polymer nanocomposites has increased exponentially with time, especially since the beginning of the 21st century, and has consistently been about 15% of the publications devoted to CNTs, as indicated in Figure 4.1[4]. Depending on end applications, CNT/polymer nanocomposites can be classified into structural and functional composites[5]. The excellent mechanical properties of CNTs, such as the high modulus, tensile strength, and strain to fracture, are explored to obtain structural composites with exceptional structural performance. Meanwhile, many interesting physical properties of CNTs, such as excellent

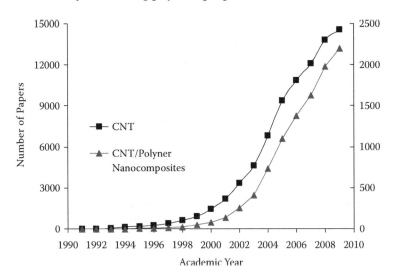

FIGURE 4.1
Number of published papers related to CNT and CNT/polymer nanocomposites as a function of academic year. Adapted from Ma, P.C. et al. 2010. Compos A 41: 1345.

electrical, thermal, optical, and damping properties, along with their excellent mechanical properties, are utilized to develop nanocomposites with multifunctional properties for applications in the fields of heat resistance, chemical sensing, electrical and thermal management, photoemission, photovoltaic films, electromagnetic absorption, and energy storage.

This chapter is devoted to studying various CNT/polymer nanocomposite fabrication techniques as well as the effects of CNT dispersion and functionalization on the properties of CNT/polymer nanocomposites. In-depth discussions are offered on the role of the CNT–polymer interface in determining the properties and mechanical performance of nanocomposites. With respect to design and service in end applications, comprehensive studies are also presented on how the CNT–polymer interface can be tailored to achieve balanced properties and desired characteristics.

4.2 Fabrication for CNT/Polymer Nanocomposites

4.2.1 Solution Mixing

Solution mixing is the most common method for fabricating CNT/polymer nanocomposites because it is amenable to small sample sizes[5–7]. Solution blending involves typically three major steps: (1) dispersion of CNTs in a suitable solvent that can also dissolve the polymer matrix by mechanical mixing, (2) mechanical/magnetic agitation or ultrasonication, (3) a mixture of the dispersed CNTs with the polymer at room or elevated temperature. The final composites are obtained by precipitating or casting the mixture using a mold. This method is also often used to prepare composite films.

4.2.2 Melt Blending

Melt blending is another commonly used method for fabricating CNT/ polymer nanocomposites. Most thermoplastic polymers, such as polypropylene[8], polystyrene[9], poly(ethylene 2,6-naphthalate)[10], can be processed as matrix materials in this method. The major advantage of this method is that no solvent is employed to disperse CNTs. Melt blending uses an elevated temperature and a high shear force to disperse CNTs in a polymer matrix, and is most compatible with current industrial practices. Special equipment, such as an extruder, injection machine, and calendering machine, which are capable of being operated at an elevated temperature and generating high shear forces, are employed to disperse CNTs. Melt blending or variants of this technique are frequently used to produce CNT/polymer composite fibers. Compared with the solution mixing method, this technique is generally considered less effective in dispersing CNTs in polymers than is solution mixing, and its application is also limited to low filler concentrations due to the high viscosities of the nanocomposites at high CNT contents[7].

4.2.3 In-Situ Polymerization

In-situ polymerization is considered to be a very efficient method of significantly improving the CNT dispersion and enhancing the interaction between CNTs and the polymer matrix. In this method, CNTs are dispersed with monomers and then the monomers are polymerized via addition or condensation reactions. One of the major advantages of this method is that covalent bonds can be formed between the functionalized CNTs and the polymer, resulting in much improved properties of the composites compared to the matrix. Epoxy-based nanocomposites comprise the majority of reports based on in-situ polymerization methods[11-16], where the nanotubes are first dispersed in the resin followed by curing of the resin with a hardener at elevated temperatures (Figure 4.2). Note that with a high content of CNTs, the viscosity of the reaction medium increases as polymerization progresses, bringing some negative effects on the curing of the thermosetting matrices.

The functionalization of CNTs using polymers via the *grafting from* method can be regarded as another kind of in-situ polymerization for the fabrication of CNT/polymer nanocomposites. The principles of this method are discussed in Chapter 3.

4.2.4 Latex Technology

A relatively new approach to incorporate CNTs into a polymer matrix is based on the use of latex technology[5-7]. Latex is a colloidal dispersion of discrete polymer particles, usually in an aqueous medium. By using this

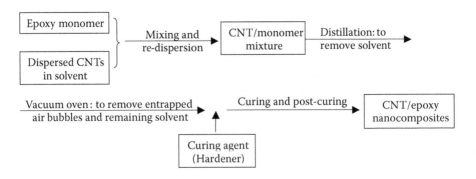

FIGURE 4.2
Schematics of CNT/epoxy nanocomposites fabrication. Adapted from Ma. P. C., et al. 2009. *ACS Appl. Mater. Interfaces* 1: 1090.

technology, it is possible to disperse single and multiwalled CNTs within most of the polymers that are produced by emulsion polymerization, or that can be brought into the form of an emulsion. Unlike the in-situ polymerization systems, the addition of CNTs in this technique takes place after the polymer has been synthesized. The whole process basically consists of simple mixing of two aqueous components. The first step consists of exfoliation of single-walled CNTs (SWCNT) bundles or dispersion/stabilization of multiwalled CNTs (MWCNT) entanglements in an aqueous surfactant solution, which is followed by mixing the stable dispersion of CNTs covered by surfactant molecules with polymer latex. After freeze drying and subsequent melt processing, a composite consisting of dispersed CNTs in a polymer matrix is obtained.

The advantages of this technique are obvious[5,6]: it is easy, versatile, reproducible, and reliable, and allows uniform incorporation of individual CNTs into a highly viscous polymer matrix. The solvent used for CNT dispersion is water, making the whole process a safe, environmentally friendly, and low-cost method. Nowadays, polymer latex is industrially produced on a large scale and this type of industry is mature. Since this process is deceptively simple and requires only a few steps, it holds the highest prospect for the scale-up fabrication of CNT/polymer nanocomposites.

4.2.5 Resin Transfer Molding

Resin transfer molding (RTM) is a process that involves placing a textile preform into a mold, injecting the mold with a liquid resin at a low injection pressure, and curing the resin to form a solid composite. It is an advanced technique for the fabrication of fiber-reinforced polymer composites. Compared with other methods, RTM is a simple process that can make composites in large sizes and complex shapes within a short cycle time and at low cost, and can be done with the majority of thermosetting resins with low viscosities.

Jiang et al. developed a method to form continuous CNT yarns or sheets by directly drawing CNTs from super-aligned CNT arrays[17–19]. These continuous and aligned CNT sheets can be stacked together to produce a CNT preform with thickness on a centimeter scale, which can be further infiltrated with a resin using RTM to fabricate CNT/epoxy nanocomposites[20]. Figure 4.3A shows the schematic and scanning electron microscopy (SEM) images of stacked CNT sheets in the same direction [0] or in two perpendicular directions alternatively [0/90]. Figure 4.3B shows the schematic of the RTM process for fabricating aligned CNT/epoxy nanocomposites from the CNT preform inserted into the RTM mold. The thickness of nanocomposites can be accurately controlled by inserting steel strips of varying thicknesses. A liquid epoxy resin was injected into the sealed RTM mold at a pressure of 0.2 MPa to overflow and fully infiltrate the CNT preform at 60°C in a vacuum oven. At this infiltration temperature, the viscosity of

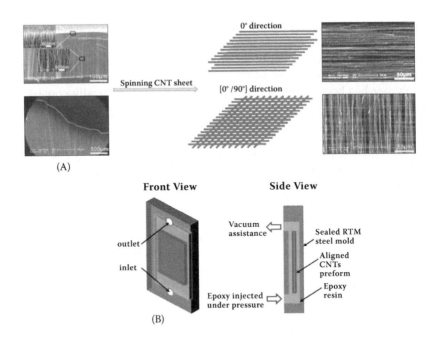

FIGURE 4.3
(A) Schematic and SEM images of CNT preforms with [0] and [0/90] alignment of CNT sheets and (B) the RTM process for fabricating CNT/epoxy nanocomposites. Adapted from Chen. Q. F., et al. 2010. *Carbon* 48: 260.

the epoxy system was kept at as low as 50 cPa·S for more than 12 hours. The temperature of the oven was subsequently increased to 120°C at a rate of 2°C/minute and held at 120°C for 12 hours to cure the epoxy, forming a solid-state aligned CNT/epoxy nanocomposite. The RTM mold was then cooled down to room temperature and the CNT/epoxy nanocomposite sample was released from the RTM mold.

The unique features of nanocomposites made by RTM are that CNTs with a weight fraction up to 16.5 wt% can be homogenously dispersed and highly aligned in the epoxy matrix. Both the mechanical and electrical properties of the nanocomposites are dramatically improved with the addition of the CNTs[20].

4.2.6 Other Methods

To obtain CNT/polymer nanocomposites with a very high CNT content or for some specific applications, new methods have been developed in recent years. Table 4.1 summarizes these methods and their features are briefly presented. It should be noted that as an emerging field, the methods for fabrication of CNT/polymer nanocomposites are not limited to those discussed herein and new techniques appear now and then.

TABLE 4.1 New Methods Developed Recently for the Fabrication of CNT/Polymer Nanocomposites.

Item / Method	CNT	Fabrication process	Advantages	Reference
Densification	As-grown CNT forest	CNT forest is prepared and transferred to a pool of uncured epoxy. The resin is infused into the CNT forest and then cured.	CNT vol% can be controlled (5–20 vol%) by varying the densification of CNT forest. Possible alignment of CNTs in nanocomposites.	21
Coagulation spin	Predispersion of CNTs using a surfactant solution	Coagulation of CNTs into a mesh by wet spinning into a polymer solution, and converting the mesh into a solid fiber by a slow draw process.	Employed to fabricate CNT/polymer fiber.	22
Layer-by-layer deposition	Predispersion of CNTs in an organic solvent	Dipping of a solid substrate like glass slides and silicon wafers into the CNT/polymer solution and curing.	Structural defects originating from the phase segregation of polymer/CNT can be minimized. High CNT concentration (up to about 50 wt%).	23
Pulverization	As-produced CNTs	Polymer and CNTs are mixed and pulverized using a pan mill or twin screws.	Possible grafting of polymers on CNTs. Easy scale-up and solventless process.	24, 25

4.3 Effects of CNT Dispersion and Functionalization on the Properties of CNT/Polymer Nanocomposites

4.3.1 Dispersion Behavior of Functionalized CNTs in Polymer Matrices

There are two major issues for an effective dispersion of nanofillers in a polymer resin: one is to disperse agglomerated bundles into individual particles (see Chapter 2); the other is to maintain stable dispersion of nanoparticles to avoid secondary agglomeration before consolidation because the nanoparticles tend to attract one another due to the van der Waals forces and Coulomb attractions to form large agglomerates[26,27]. Many previous studies confirmed beneficial effects of CNT functionalization on dispersion and long-term stability of CNTs in aqueous solutions or in polymer resins[11,13,14,27–35]. For example, Shofner et al.[33] reported ameliorating effects of sidewall functional groups on dispersion of fluorinated SWCNTs in a polyethylene (PE) matrix as well as on the mechanical properties of the corresponding nanocomposites. In addition, the observed partial removal of functional groups from the fluorinated SWCNTs during melt processing with PE suggested a possibility of in situ direct covalent bonding between the nanotubes and the matrix, giving rise to mechanical reinforcement of the composite. Chen et al.[34] functionalized CNTs with polyhedral oligomeric silsesquioxane (POSS) via amide linkage. The energy-filtering transmission electron microscopy (TEM), Fourier transform infrared (FT-IR) spectroscopy, and Raman spectroscopy revealed that the POSS was covalently attached to the MWCNTs and the weight gains arising from the functionalization were reflected by the thermogravimetric analysis (TGA) result.

An elevated temperature is generally required for curing of the majority of thermosetting matrices. The elevated temperature may also serve as a driving force to reactivate the CNTs to reagglomerate as the viscosity of resin is significantly reduced at the initial stage of the curing process. It is unknown, however, whether the functionalization really helps maintain the stability of dispersed CNTs in a polymer resin. In an effort to shed light on the above intriguing question, Ma et al. studied[35] the dispersion stability of amino-functionalized CNTs (amino-CNTs) during curing. A liquid epoxy resin containing 0.05 wt% of CNTs was subjected to a curing temperature of 80°C on a hot stage while the dispersion state of CNTs were being monitored under an optical microscope. Figure 4.4 shows the change in dispersion state of CNTs at different stages of curing. Both properly dispersed particles and agglomerates were seen in the epoxy matrix containing the pristine CNTs (P-CNTs) before curing (Figure 4.4A). The dispersion state of P-CNTs in the matrix exhibited obvious changes with time, in particular at the very early stage of curing, indicating a flow of resin. The P-CNTs were agglomerated within 2 minutes of curing (see Figures 4.4A–C), whereas the matrix-rich

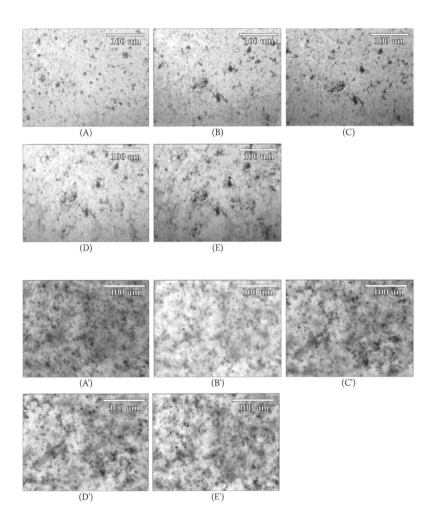

FIGURE 4.4
Optical micrographs of P-CNTs (A to E) and amino-CNTs (A′ to E′) with a constant CNT content of 0.05 wt% in an epoxy resin, taken at different curing times of 0, 1, 2, 5, and 60 minutes (from left to right). Adapted from Ma, P. C., et al. 2010. *Carbon* 48: 1824.

regions, that is, the transparent regions, became more pronounced with increasing curing time (see Figures 4.4C–E). All these observations clearly suggest reagglomeration of P-CNTs during curing, which is thought to be a time-dependent process. It is expected that the reagglomeration can be more pronounced at higher P-CNT contents because the interparticle distances between dispersed CNTs are reduced significantly.

In contrast, the CNT dispersion in the nanocomposite containing amino-CNTs remained fairly uniform and little changed regardless of curing time (see Figures 4.4A′–E′), confirming that functionalization indeed facilitated the stability of CNT dispersion even at a high temperature. The mechanism behind this observation is that the presence and nature of amino-functional groups attached on CNTs affect the rheological behavior of epoxy[36,37], resulting in faster curing and polymer wrapping of CNTs at the early stage of curing. As the curing proceeded, the wrapped polymer acted as a protective layer to prevent dispersed CNTs from attracting one another, thus effectively discouraging the migration of CNTs to reagglomerate.

The fracture surfaces of fully cured nanocomposites were also examined, as shown in Figure 4.5, to further confirm the previous observation. Large CNT agglomerates were clearly seen from the nanocomposites containing P-CNTs (Figure 4.5A), resembling the optical images taken after 60 minutes (Figure 4.4E). These agglomerates were almost absent in the nanocomposites containing amino-CNTs (Figure 4.5C), confirming that (1) the dispersion of CNTs was enhanced by functionalization, and (2) the improved dispersion of CNTs was stabilized and maintained after the epoxy resin was fully cured. In addition, a close examination of the SEM images indicated many voids and long pull-out tubes on the nanocomposites containing P-CNTs (Figure 4.5B),

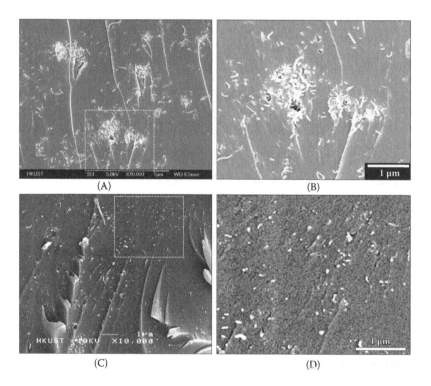

(A)   (B)

(C)   (D)

FIGURE 4.5
SEM images of fracture surface of nanocomposites containing different CNTs: (A and B) 0.5 wt% P-CNTs, and (C and D) 0.5 wt% amino-CNTs. Adapted from Ma, P. C., et al. 2010. *Carbon* 48: 1824.

suggesting weak CNT–matrix adhesion. In contrast, the pull-out CNTs were relatively short and in a small number (Figure 4.5D) because the functionalized CNTs adhered strongly to the matrix due to the covalent interactions.

4.3.2 Mechanical Properties

CNTs possess excellent mechanical properties with a Young's modulus as high as 1.2 TPa and a tensile strength of 50–200 GPa. The combination of these exceptional mechanical properties along with the low density, high aspect ratio, and high surface area make CNTs ideal candidates for reinforcement in composite materials. Both SWCNTs and MWCNTs have been utilized for reinforcing thermosetting polymers, such as epoxy, polyurethane, and phenol-formaldehyde resins, as well as thermoplastic polymers, including polyethylene, polypropylene, polystyrene, and nylon. The CNT-reinforced nanocomposites can be considered a kind of particulate composite with the filler dimensions on the nanometer scale and the aspect ratio as high as 1000. Unlike the microscale particulate composites, the mechanical properties of nanocomposites depend strongly on the dispersion state of nanofillers, apart from the properties of fillers and matrix material. In addition to dispersion, there are other major requirements that need to be satisfied for effective reinforcement of CNTs in composites[38]. These requirements include a high aspect ratio, preferential alignment of CNTs in the loading direction, and strong interfacial interactions between the CNT and polymer matrix.

The aspect ratio must be sufficiently high to maximize the load transfer between the CNTs and the matrix material, thus enhancing the mechanical properties. For example, polystyrene composites reinforced with well-dispersed 1.0 wt% CNTs of a high aspect ratio had more than 35% and 25% increases in elastic modulus and tensile strength, respectively[39]. Similar encouraging results have been reported with many different matrix materials (see, for example,[40,41]), while other reports demonstrated only modest improvements in modulus and strength. For example, the results of CNT/epoxy nanocomposites reinforced by two types of MWCNTs with different aspect ratios showed significantly improved impact resistance and fracture toughness only for those containing CNTs of a higher aspect ratio[42]. However, the corresponding tensile modulus and strength showed very limited improvements of less than 5%, probably due to weak bonds between the CNTs and polymer matrix and agglomeration of CNTs.

Indeed, dispersion is the most important issue in producing CNT/polymer nanocomposites. Many different techniques have been employed for CNT dispersion, as discussed in Chapter 2. A good dispersion not only makes more filler surface area available for bonding with the polymer matrix, but also prevents the aggregated fillers from acting as stress concentrators, which are detrimental to mechanical performance of composites[43]. However, some challenges, such as CNT content in composites, length and entanglement of CNTs, as well as viscosity of the matrix have to be overcome to obtain uniform CNT dispersion in nanocomposites. Many reports[13–16] showed that there was a critical CNT content in the matrix below which the strength modulus of CNT/polymer composites increased with increasing CNT content. Above this critical CNT content, however, the mechanical properties of CNT/polymer composites decreased, and in some cases, they decreased even below those of the neat matrix material. These observations can be attributed to several factors: (1) the difficulties associated with uniform dispersion of CNTs at high CNT contents and (2) lack of polymerization reactions that are adversely affected by the high CNT content. The latter effect becomes more pronounced when functionalized CNTs are employed to produce CNT/polymer nanocomposites.

The technique employed for CNT dispersion can influence, to a large extent, the mechanical properties of CNT/polymer nanocomposites. Table 4.2[44] summarizes the flexural properties of CNT/epoxy nanocomposites fabricated from the same CNT type (NK-50, NanoKarbon), the same content of 0.5 wt%, and the same epoxy matrix using different techniques for CNT dispersion. Examinations were made on polished thin nanocomposite specimens[44] as shown in Figure 4.6. The best CNT dispersion was achieved by employing a calendering machine (Figure 4.6D), and the nanocomposites fabricated using this technique exhibited the best flexural properties among the five techniques used (Table 4.2).

From a geometric standpoint, the difference between random orientation and alignment of CNTs can result in significant changes in various properties of composites[38]. For example, it was shown that the storage modulus of the polystyrene composite films containing random and oriented CNTs were 10% and 49% higher than the unreinforced bulk polymer, respectively[45]. The alignment can be regarded as a special case for the CNT dispersion state. A few techniques, including mechanical stretching, melt spinning, dielectrophoresis, and application of an electrical or magnetic field, have been employed during composite fabrication to align CNTs in a polymer matrix. The degree of CNT alignment in the composite is in general governed by two factors:

TABLE 4.2 Effect of CNT Dispersion on the Flexural Properties of CNT/Epoxy Nanocomposites Containing the Same Type and Content of CNTs.

CNT Dispersion Technique	Flexural Modulus (GPa)	Flexural Strength (MPa)
Neat epoxy	3.43 ± 0.01 (+0.00%)	140.0 ± 2.3 (+0.00%)
Sonication in water bath	3.41 ± 0.04 (–0.58%)	144.1 ± 1.8 (+2.93%)
Shear mixing	3.36 ± 0.09 (–2.08%)	140.4 ± 2.4 (+0.29%)
Probe sonication	3.47 ± 0.06 (+1.17%)	142.7 ± 1.7 (+1.93%)
Calendering	3.65 ± 0.02 (+6.41%)	145.2 ± 0.82 (+3.71%)

Source: Adapted from Ma, P. C., et al. 2008. Effect of dispersion techniques and functionalization of CNTs on the properties of CNT/epoxy composites. Paper presented at the 2nd International Conference on Advanced Materials and Structures (ICAMS-2), Nanjing, China, October 2008.

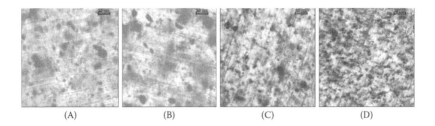

(A)	(B)	(C)	(D)

FIGURE 4.6

Dispersion of CNTs in nanocomposites using different techniques: (A) sonication in water bath, (B) shear mixing, (C) probe sonication, and (D) calendering. Scale bar = 20 μm. Adapted from Ma, P. C., et al. 2008. Effect of dispersion techniques and functionalization of CNTs on the properties of CNT/epoxy composites. Paper presented at the 2nd International Conference on Advanced Materials and Structures (ICAMS-2), Nanjing, China, October 2008.

(1) the diameter of the CNTs and (2) CNT content. A smaller diameter CNT can enhance the degree of CNT alignment due to the greater extensional flow, while alignment becomes more difficult with increasing CNT content because of CNT agglomeration and restrictions in motion caused by neighboring CNTs[46]. While alignment is necessary to maximize the strength and modulus in a certain direction, it is not always beneficial because the aligned CNT composites tend to have highly anisotropic properties, that is, the mechanical properties along the alignment direction can be enhanced, whereas these properties are sacrificed along the direction perpendicular to this orientation.

As for fiber-reinforced composites, one of the most important requirements for a CNT nanocomposite for any structural application is that the external load applied to the composite is efficiently transferred to the nanotubes, allowing them to take up the major share of the load[13,38]. Load transfer depends on the interfacial bond between the CNT and the matrix, in much the same way as in fiber composites. If the composite is subjected to a load and perfect bonding exists between the CNTs and the matrix, then the load can be transferred to the CNTs. Since the strength of CNTs is much higher than that of the matrix, the composite can withstand high loads, resulting in improved mechanical properties. In contrast, if the CNT–polymer interfacial bond is weak, then the interface would debond prematurely, well before transferring the load to the CNTs; thus the composites exhibit a lower strength. Many chemical and physical functionalization techniques have been shown to be efficient in enhancing interfacial interactions between CNTs and various polymers (see Chapter 3).

Table 4.3 summarizes recent studies about the effects of CNT functionalization on the mechanical properties of CNT nanocomposites made from thermosets and thermoplastics. These results clearly indicate that functionalization of CNTs can enhance the modulus strength as well as fracture resistance of nanocomposites, among the many other mechanical properties that were measured.

TABLE 4.3   Effect of CNT Functionalization on the Mechanical Properties of CNT/Polymer Nanocomposites.

Matrix		Dispersion Technique	Functionalization Technique	CNT Content	Enhancement in Mechanical Properties[a]			Reference
					Modulus	Strength	Toughness	
Thermo-plastic[b]	PA	Twin-screw extruder	Diamine treatment	1.0 wt%	6.1% (42%)	-5.3% (18%)	—	47
	PB	Shear mixing	Polymer grafting	1.5 wt%	18% (91%)	-27% (61%)	-40% (67%)	48
	PE	Shear mixing	Maleic anhydride and amine treatment	1.5 wt%	22% (75%)	-17% (33%)	-69% (61%)	49
	PI	Ultrasonication	Acid treatment	7.0 wt%	39% (61%)	19% (31%)	—	50
	PP	Ultrasonication and stir	Undecyl radicals attachment	1.5 wt%	55% (84%)	-10% (13%)	—	51
	PS	Probe ultrasonication	Butyl attachment	0.25 wt%	-8.3% (25%)	2.1% (50%)	—	52
	PVA	Ultrasonication and stir	Polymer grafting	2.5 wt%	35% (40%)	-4.8% (17%)	—	53
	PMMA	Ultrasonication	Polymer grafting	0.10 wt%	57% (104%)	-2.7% (86%)	—	54
Thermo-set[c]	EP	Calendering	Amino treatment	0.10 wt%	2.1% (6.7%)	-2.2% (3.1%)	17% (19%)	11
		Probe ultrasonication	Surfactant treatment	0.25 wt%	8.6% (24%)	6.8% (20%)	35% (60%)	55
		Ultrasonication	Organic silane	0.25 wt%	8.7% (22%)	3.6% (18%)	-22% (8.5%)	13
		Ultrasonication	Amino treatment	0.50 wt%	8.6% (21%)	4.9% (22%)	—	35
	PU	Ball mill	Polymer grafting	0.7 wt%	48% (178%)	27% (23%)	—	56
		Ultrasonication	Acid treatment	10 wt%	340% (500%)	51% (111%)	—	57
	VR	Ball mill	Acid treatment	25 phr	444% (594%)	175% (244%)	—	58
		Roll mill	Organic silane	1 phr	35% (28%)	-7.8% (25)	—	59

[a] Data in parentheses indicate the enhancement on mechanical properties by employing functionalized CNTs.

[b] PA (nylon): polyamide; PB: polybutylene; PE: polyethylene; PI: polyimide; PP: polypropylene; PS: polystyrene; PVA: poly (vinyl acetate); PMMA: poly(methyl methacrylate).

[c] EP: epoxy; PU: polyurethane; VR: vulcanized rubber.

Fracture toughness is an important material property for many engineering applications because it reflects the ability of a material to resist crack propagation and fracture. As shown in Table 4.3, the incorporation of functionalized CNTs into a polymer matrix also enhances the fracture toughness of nanocomposites. Ma et al.[13] measured the quasi-static fracture toughness, K_{IC} of epoxy-based nanocomposites filled with silane-functionalized CNTs and untreated CNTs, and found that the general trends of K_{IC} with respect to CNT content were largely different depending on whether the CNT was functionalized or not (Figure 4.7). The addition of untreated CNTs into epoxy resulted in a gradual reduction of K_{IC}, whereas the nanocomposites containing silane-functionalized CNTs (silane-CNTs) showed a moderate increase in K_{IC}. These observations can be explained in terms of dispersion and interfacial interactions between the CNT and the epoxy. For the untreated CNTs, both the dispersion in epoxy and the interfacial interaction were poor due to the agglomeration and the inherently inert/ hydrophobic nature of CNTs. After silane functionalization, the dispersion of CNTs and the interfacial interaction with epoxy was enhanced through the attachment of oxygen-containing functional groups and silane molecules onto the CNT surface[13].

Furthermore, major fracture mechanisms were also identified based on the morphological analysis of the fracture surfaces of nanocomposites taken near the initial crack tip, as shown in Figures 4.8 and 4.9[13]. The neat epoxy (Figure 4.8A) exhibited a smooth, mirror-like fracture surface, representing brittle failure of a homogeneous material. No obvious difference was observed between the neat epoxy and the composite containing 0.05 wt% untreated

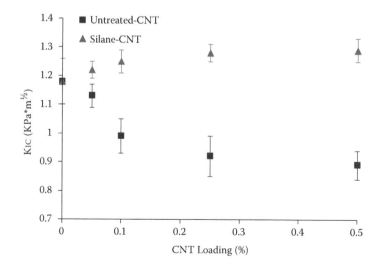

FIGURE 4.7
Fracture toughness, K_{IC}, of CNT/epoxy nanocomposites with different CNT functionalities and contents. Adapted from Ma. P. C., et al. 2007. *Compos. Sci. Technol.* 67: 2965.

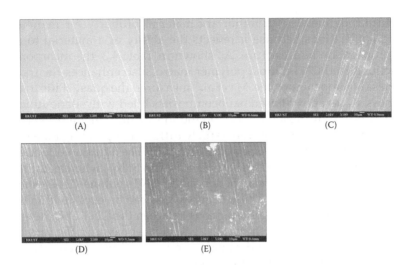

FIGURE 4.8
Fracture surface morphologies of nanocomposites containing untreated CNTs: (A) neat epoxy, (B) 0.05 wt% CNTs, (C) 0.1 wt% CNTs, (D) 0.25 wt% CNTs, and (E) 0.5 wt% CNTs. Adapted from Ma. P. C., et al. 2007. *Compos. Sci. Technol.* 67: 2965.

CNT (Figure 4.8B). With further increase in CNT content (Figures 4.8C–E), the CNTs tended to agglomerate into larger sizes, while the overall fracture surface morphology did not alter much with only marginal increases in the number of river markings. However, the corresponding fracture toughness consistently decreased because the CNT aggregates acted as the stress concentrators.

In sharp contrast, the fracture surfaces for the composites containing silane-CNTs revealed a systematic increase in the number of river markings and the corresponding surface roughness with increasing CNT content. At low CNT contents (0.05–0.1 wt%), there were straight and long river markings running parallel to the crack propagation direction (Figures 4.9A and B). At higher CNT contents (0.25–0.5 wt%, Figures 4.9C–D), these river markings became shorter and round-ended, similar to those observed in nanocomposites reinforced with high contents of nanoclay particles[60]. It appears that the increasing number of river markings roughly corresponds to the number of isolated, well-dispersed CNTs, which forced the cracks to propagate, bypassing the CNTs and taking longer paths. This resulted in dissipation of more energy through the well-known pinning and crack tip bifurcation mechanisms. However, the fracture toughness became more or less saturated at about 0.5 wt% CNT content, indicating that a higher CNT content would not result in further increase in fracture resistance because the chance of agglomeration increases with increasing CNT content.

It should be noted that an excessive degree of CNT functionalization will not only damage the structure of the CNTs, but also adversely affect the curing reactions of thermosetting matrices, resulting in negative effects

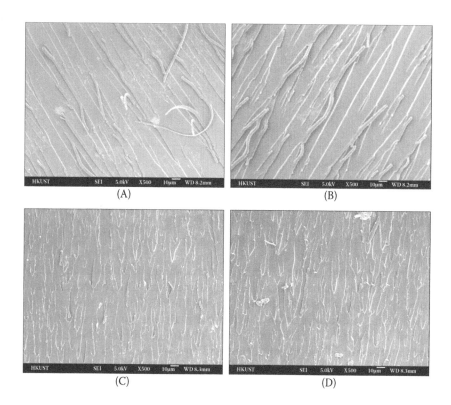

FIGURE 4.9
Fracture surface morphologies of nanocomposites containing silane-CNTs: (A) 0.05 wt% CNTs, (B) 0.1 wt% CNTs, (C) 0.25 wt% CNTs, and (D) 0.5 wt% CNTs. Adapted from Ma. P. C., et al. 2007. *Compos. Sci. Technol.* 67: 2965.

on mechanical properties of CNT/polymer nanocomposites. Figure 4.10 shows the elastic moduli and strengths of nanocomposites filled with pristine CNTs (P-CNTs), silane-CNTs, and amino-CNTs[35]. The nanocomposites containing functionalized CNTs (i.e., both silane-CNTs and amino-CNTs), exhibited a more pronounced enhancement in modulus over the whole CNT contents studied than the P-CNT counterpart. The flexural strengths of the composites presented basically a similar trend: the strengths of the composites containing functionalized CNTs were consistently higher than the P-CNT counterpart. The improvements in both modulus and strength arose from the combination of better CNT dispersion and stronger filler–matrix adhesion due to functionalization. It is worth mentioning that the nanocomposites containing silane-CNTs exhibited a higher flexural modulus than those containing amino-CNTs, whose difference diminished at a higher CNT content, say 0.5 wt% CNTs. The flexural strengths were initially similar between the two functionalized CNTs, but the composites containing amino-CNTs outperformed when the CNT content was 0.5 wt%. In summary,

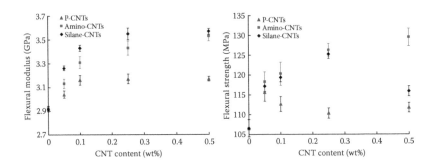

FIGURE 4.10
Flexural properties of nanocomposites with different CNT functionalities and contents. Adapted from Ma, P. C., et al. 2010. *Carbon* 48: 1824.

the amino-CNTs were more effective in enhancing the mechanical properties of composites, particularly at high CNT contents. The reason is that the amino-functionalization tended to damage the CNT structure much less than silane functionalization, thus better maintaining the inherent structure and properties of CNTs. High-resolution TEM images of the CNTs with different functionalities indicate that the end tips of the P-CNTs were closed before any functionalization process (Figure 4.11A). The sound graphene structure of the amino-CNTs (Figure 4.11B) confirmed little damage to the CNT surface. However, the end tips of the silane-CNTs were open or curved (Figure 4.11C), a reflection of broken C–C bonds along the graphene layers of the coaxial tubes. Another reason is that the wrapping of silane molecules on the CNT surface affected the stoichiometry of the curing reaction between the diglycidyl ether of bisphenol-A (DGEBA) epoxy and amine hardener, resulting in adverse effects on both the cross-linking reaction and uniform dispersion of CNTs, especially at high CNT contents.

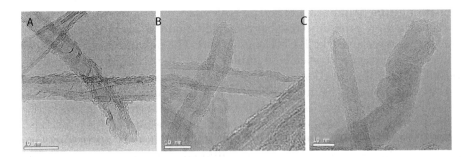

FIGURE 4.11
High-resolution TEM images of CNTs with different functionalities: (A) P-CNTs, (B) amino-CNTs, and (C) silane-CNTs. Adapted from Ma, P. C., et al. 2010. *Carbon* 48: 1824.

4.3.3 Electrical Properties

CNTs have been employed as fillers in a polymer matrix to fabricate conducting nanocomposites. Electrically conducting composites, consisting of conducting fillers in an insulating polymer matrix with a volume conductivity higher than 10^{-10} S/cm, are considered to be an important group of relatively inexpensive materials useful for many engineering applications (Figure 4.12A)[6,61,62], such as electrically conducting adhesives, antistatic coatings and films, electromagnetic interference shielding materials for electronic devices, thermal interface materials, and so on. The percolation theory was applied to explain the electrically conducting behavior of composites containing conducting fillers and insulating matrices. The relationship between the filler content and electrical conductivity of composites can be described by the following power law equation[63–65]:

$$\sigma = \sigma_0 (V_f - V_c)^s \tag{4.1}$$

where σ is the electrical conductivity of a composite, σ_0 is the electrical conductivity of the filler, V_f is the filler volume fraction, Vc is the percolation threshold, and s is a conducting exponent. When the conducting filler content is gradually increased, the composite undergoes an insulator-to-conductor transition and a typical percolation behavior can be observed. A critical filler content is referred to as the percolation threshold, V_c, where the measured electrical conductivity of the composite sharply jumps up several orders of magnitude due to the formation of continuous electron paths, or a conducting network, in the insulating polymer matrix. Below the percolation transition range, electron paths do not exist, and the electrical properties are dominated by the matrix material. Above the percolation transition range, multiple electron paths exist in the matrix, thus the electrical conductivity of the composite reaches a saturation value. This behavior is graphically shown in Figure 4.12B[63–65].

The concentration of conducting fillers must be maintained above the percolation threshold in order to form conducting networks in conducting composites. Conventional conducting fillers are usually micrometer-scale metal powder or carbonaceous materials, such as carbon black, exfoliated graphite, or carbon fibers. To achieve satisfactory electrical conductivity, the conventional filler content needs to be as high as 10 to 50 wt%[62–66], resulting in a composite with poor mechanical properties and high density. The use of CNTs for producing conducting composites can minimize these problems. Compared with traditional conducting fillers, CNTs have unique advantages of high aspect ratios and excellent electrical conductivities, which facilitate the formation of conducting networks in composites and transportation of electrons along the networks, thus making the insulating polymer into a conducting composite with a very low CNT content.

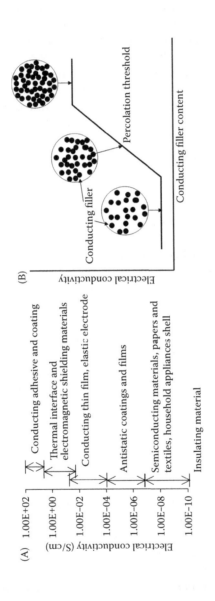

FIGURE 4.12

(A) Typical applications of conducting composites, and (B) schematics of percolation phenomenon and conducting network in conducting composites. Adapted from Aneli, J. N., et al. 1998. *Structuring and conductivity of polymer composites*, 50. Commack, NY: Nova Science; Li, J. 2006. Electrical conducting polymer nanocomposites containing graphite nanoplatelets and carbon nanotubes. PhD thesis. Hong Kong University of Science and Technology 1–100; and Ma, P. C. 2008. Novel surface treatment, functionalization and hybridization of carbon nanotubes and their polymer-based composites. PhD thesis. Hong Kong University of Science and Technology 1–30.

Figure 4.13 summarizes the percolation threshold of nanocomposites consisting of different types of polymer matrix and CNT[67]. The percolation threshold values of most polymer matrices were below 5 wt%, with the exception of a few isolated cases of polyamide (PA), and poly (vinyl alcohol) (PVA) matrices. Nevertheless, there was no apparent consensus on percolation thresholds even for apparently the same type of matrix: for example, the values reported for typical CNT/epoxy nanocomposites varied from 0.002 to over 7 wt%[12,13,68,69], depending on the type of CNTs (SWCNTs or MWCNTs) and the processing techniques used to produce the nanocomposites. The large variation indicates that the dispersion states and the properties of CNTs affected by functionalization and processing conditions are important in determining the electrical properties of CNT/polymer nanocomposites. Bauhofer et al.[67] summarized the effect of experimental parameters on the percolation threshold of CNT/polymer nanocomposites, and concluded that well-dispersed CNTs gave conductivities 50 times higher than the entangled ones, and the CNT dispersion played a more important role in determining the electrical behavior of CNT/polymer nanocomposites than the type and production method of CNTs.

Li et al.[27] produced CNT/epoxy nanocomposites by employing four different techniques for CNT dispersion: (A) shear mixing, (B) ultrasonication, (C) ultrasonication and shear mixing, and (D) ball milling and ultrasonication. It was found that the electrical conductivity varied significantly depending

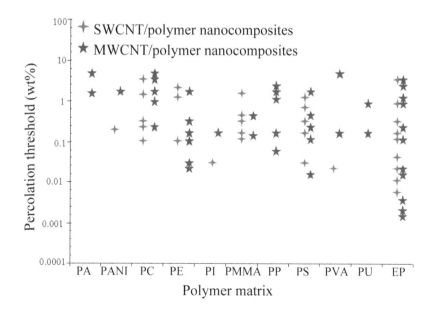

FIGURE 4.13
Percolation threshold of CNT/polymer nanocomposites. Adapted from Bauhofer, W., et al. 2009. *Compos. Sci. Technol.* 69: 1486.

on the processing conditions for the nanocomposites. The percolation threshold of the nanocomposite prepared by Condition A was high (0.4 wt%) (Figure 4.14A) because the large-sized, tightly entangled CNT agglomerates made it difficult to form conducting networks with low filler contents (Figure 4.14B). Condition B had the lowest percolation threshold of 0.1 wt% (Figure 4.14A) because of the loosely entangled and uniformly distributed CNT agglomerates (see Figure 4.14B). A moderate percolation threshold value was observed in Condition C (0.25~0.3 wt%; Figure 4.14A) with satisfactory CNT dispersion (Figure 4.14B). Although CNTs in Condition D exhibited the best dispersion state on the nanoscopic scale (Figure 4.14B), no apparent percolation threshold was observed even with 1 wt% CNT content (Figure 4.14A). This is due to severe damage of CNTs during the combined ball milling and ultrasonication treatment, resulting in a large reduction in the CNT aspect ratio. They further concluded that the critical factors determining the percolation threshold of CNT/polymer nanocomposites were (1) the aspect ratio of the CNTs, (2) disentanglement of CNT agglomerates on the nanoscopic scale, and (3) uniform distribution of individual CNTs or CNT agglomerates on the microscopic scale. There was a critical CNT aspect ratio value, above which the dispersion of CNTs became crucial, allowing the percolation threshold to vary by several orders of magnitude, while below this value the percolation threshold increased rapidly with a decreasing aspect ratio[27].

Many studies have shown that CNT functionalization has tremendous influence on the electrical conductivities of nanocomposites. Proper functionalization facilitates CNT dispersion and the formation of conducting networks in composites, resulting in lowered percolation thresholds of nanocomposites. However, excessive functionalization introduces too many heterogeneous atoms on the CNT surface, resulting in the perturbation of π electrons and thus degrading the intrinsic electrical properties of CNTs. In addition, functionalization of CNTs using acids in high concentrations can severely damage and fragment CNTs into smaller pieces with decreased aspect ratios. Both of these are detrimental to the formation of conducting networks in nanocomposites. Figure 4.15 shows the effect of CNT functionalization on the electrical conductivities of CNT/epoxy nanocomposites. The CNTs functionalized with an organic silane led to wrapping of insulating material on the CNTs[13], the wrapping being more serious for well-dispersed individual CNTs, thus exhibiting no percolation behavior for the nanocomposites containing the silane-treated CNTs. When amino-functionalized CNTs were employed[14], the nanocomposites showed a typical percolation behavior, although the conductivities were in general lower than those containing pristine CNTs.

The electrical conductivities of CNT/polymer composites typically ranged from 10^{-5} to 10^{-3} S/cm above the percolation threshold at CNT contents below about 0.5 wt%[13,14,68,69]. While a further increase in CNT content above the percolation threshold can marginally enhance the electrical conductivity

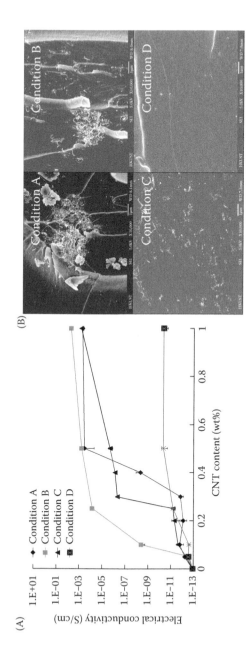

FIGURE 4.14

(A) Electrical conductivities of CNT/epoxy nanocomposites fabricated by different dispersion techniques, and (B) corresponding CNT dispersion in the matrix. Adapted from Li, J., et al. 2007. *Adv. Funct. Mater.* 17: 3207.

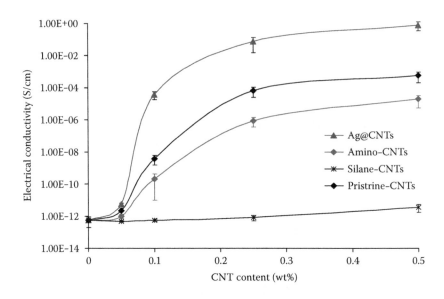

FIGURE 4.15
The effect of CNT functionalization on the electrical conductivities of CNT/epoxy nanocomposites. Adapted from Ma. P. C., et al. 2007. *Compos. Sci. Technol.* 67: 2965; and Ma. P. C., et al. 2008. *Carbon* 46: 1497.

of composites, the solution viscosity becomes too high to produce void-free composites when the CNT content is higher than 1.0 wt%. This limits the use of CNT/polymer composites for applications where high CNT contents are necessary. Therefore, processing techniques for improving the electrical conductivity of composites below or near the percolation threshold become crucial in producing highly conducting composites. Simulation results indicated that the contact resistance of CNTs in composites played an important role in enhancing the conductivity of nanocomposites[70,71]. This was confirmed by a recent study[14]; when silver-decorated CNTs were employed as conducting filler, the nanocomposites exhibited a significantly higher conductivity above the percolation threshold compared with those containing pristine CNTs, as shown in Figure 4.15. A high conductivity value of 0.81 S/cm was achieved with 0.5 wt% of silver-decorated CNTs. This observation was due to the silver nanoparticles that were tightly attached to the defect sites on the CNT surface, which compensated for the negative effect of CNTs (due to amino functionalization) by enhancing the conductivity of CNTs and reducing the contact resistance of CNT junctions in the matrix.

The high cost of CNTs, especially SWCNTs, compared with other fillers, like graphite, carbon black, and carbon fibers, limits the wider applications of CNT-based conducting composites. Therefore, nanocomposites containing hybrid fillers of CNTs and lower-cost materials were developed in recent years[72-77]. In these composites, the CNT network acts as a skeleton, whereas the conducting networks formed by other fillers facilitate the transportation of electrons along the CNT skeleton; thus, both conducting fillers offer a synergy in enhancing the electrical properties of hybrid nanocomposites. Li et al.[72] reported the electrical performance of conducting nanocomposites consisting of hybrid fillers of graphite nanoplatelets (GNPs) and CNTs. The nanocomposites containing 1% GNP and 1% CNT achieved the highest electrical conductivity of 4.7×10^{-3} S/cm, more than two orders of magnitude higher than that of the nanocomposites with 2 wt% GNP alone.

In another study[16], nanocomposites reinforced with hybrid fillers of CNTs and carbon black (CB) were prepared, aiming at enhancing the electrical conductivity of composites with balanced cost and mechanical properties. The results showed enhanced electrical conductivity of hybrid composites: a remarkably low percolation threshold of 0.4 wt% was achieved with hybrid fillers of 0.2 wt% CNTs and 0.2 wt% CB particles. Furthermore, the incorporation of CB particles into the neat epoxy or CNT nanocomposites offered additional benefits of enhancing the ductility of the final product. Figure 4.16 shows typical stress-strain curves of nanocomposites containing either single fillers or hybrid fillers. The CNT nanocomposites (curves B and C in Figure 4.16A) showed higher flexural strength than the neat epoxy (curve A in Figure 4.16A) before catastrophic failure. When CB was introduced into the epoxy, the nanocomposites (curves D, E, and F in Figure 4.16A) consistently showed much larger deformation than those without CB. This behavior was most pronounced in the composites containing hybrid fillers of CNTs and CB (curve F in Figure 4.16A). These observations strongly suggested that (1) the CB particles played an important role in changing the fracture behavior of nanocomposites from brittle to ductile failure, and (2) hybridization of CNTs with CB particles improved the fracture resistance, enhancing the energy dissipation capacity of nanocomposites.

The above suggestions were verified by measuring the Charpy impact fracture toughness of the nanocomposites, as illustrated in Figure 4.16B. The nanocomposites showed a substantial increase in impact fracture toughness, especially those containing both CNT and CB particles. The nanocomposite containing hybrid fillers of 0.2% each of CNT and CB exhibited a remarkable improvement in impact toughness of more than 55% (from 4.82 to 7.57 KJ/m^2), which was the highest among the materials with a total filler content of 0.4%, further confirming the synergistic effect of hybrid fillers in enhancing the fracture resistance of nanocomposites.

4.3.4 Thermal Properties

The excellent thermal properties of CNTs, such as high thermal conductivity and good thermal stability, lead to expectations that these materials can be employed as functional fillers to rectify the thermal properties of CNT/ polymer nanocomposites. Nevertheless, the enhancements in thermal conductivity, thermal stability, and thermomechanical properties due to the

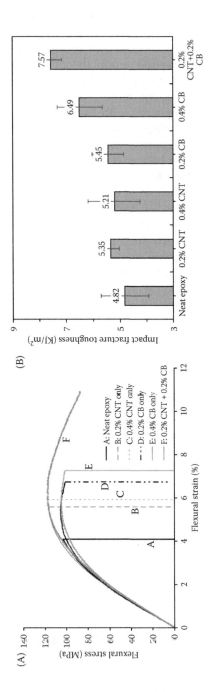

FIGURE 4.16

Improved mechanical properties of polymer nanocomposites filled with hybrid fillers of CNTs and CB: (A) stress–strain curves of nanocomposites containing different fillers, and (B) impact fracture toughness of nanocomposites containing different fillers. Adapted from Ma. P. C., et al. 2009. *ACS Appl. Mater. Interfaces* 1: 1090.

introduction of CNTs in a polymer composite have not been as impressive as expected. There are several major concerns in this area, which will be specifically discussed in the following text.

The thermal conductivity of a material is determined by atomic vibration or phonons, and the conduction by electrons is generally negligible for insulating materials[78]. For CNT/polymer composites, it critically depends on the content, aspect ratio, dispersion state of the CNTs, and the interfacial interactions between the CNTs and the polymer matrix. Much effort has been directed toward utilizing CNTs as thermal conducting filler in polymer composites and some enhancements were indeed observed. For example, Biercuk et al.[79] used SWCNTs to augment the thermal transport properties of epoxy, showing 70% and 125% increases in thermal conductivity of epoxy with a constant 1.0 wt% of SWCNTs at 40 K and room temperature, respectively. Evseeva et al.[80] found that the introduction of 0.1–1.0 wt% MWCNTs enhanced the thermal conductivity of a pure epoxy resin by about 40%. However, judging from the available reports published so far in open media, it should be noted that the thermal conductivities of the vast majority of CNT/polymer composites studied showed only a marginal improvement when compared with that of electrical conductivity. It is not unusual to see an improvement of electrical conductivity of polymer composites of as much as eight to ten orders of magnitude with less than 0.5 wt% CNT content (see, e.g., Figures 4.14 and 4.15). Taking into account the different transport mechanisms between the thermal and electrical conductivities of composites, Moniruzzaman et al.[7] proposed that the phonons, major carriers for thermal conduction, were much more likely to travel through the matrix than through the CNT networks. It should be noted that the difference in thermal conductivity between the CNT and polymer, that is, about 10^4 W/(m·K) with $K_{CNT} \approx 10^3$ and $K_{polymer} \approx 10^{-1}$ W/(m·K), is much smaller than that of the electrical conductivity, that is, in the range of $10^{15} - 10^{19}$ with $\sigma_{CNT} \approx 10^2 - 10^6$ and $\sigma_{polymer} < 10^{-13}$ S/cm.

The aspect ratio of CNTs also has a significant effect on thermal conductivity of composites. This can be understood from two viewpoints: (1) for a given volume fraction, the number of CNT-polymer-CNT contacts increases with decreasing aspect ratio, which in turn reduces the influence of high interfacial thermal resistance between the CNT and polymer matrix[81]; and (2) an increase in aspect ratio of CNTs shifts the phonon dispersion toward the lower frequencies, resulting in a better CNT–liquid thermal coupling[82]. Cai's work[83] on CNT/polyurethane composites fabricated based on latex technology has proven the effect of aspect ratio: the thermal conductivity of the composites increased by more than 200% with 3 wt% CNTs whose aspect ratio was maintained over 3000, that is, 50 μm in length and 8–15 nm in diameter.

There is another important factor that may affect the thermal conductivity, namely the interfacial adhesion between CNTs and polymer matrix. The improved CNT–polymer interfacial interactions can inhibit the phonon transportation along CNTs, and increase the interfacial thermal resistance by affecting the damping behavior arising from the vibration of phonons[84]. Ma et al.[14] compared the thermal conductivities of composites filled with CNTs with different surface functionalities (Figure 4.17), showing that the thermal conductivity was in general higher for those containing pristine CNTs than those with amino-functionalized CNTs that possessed amine and amide functional groups on the surface. They also proposed a method to recover the thermal conductivity of CNTs by decorating silver nanoparticles onto the CNT surface (see Figure 4.17). Silver is an excellent thermal conductor ($K_{Ag} = 429W/(m \cdot K)$) and the Ag–CNT interface facilitated phonon conduction by reducing the boundary scattering losses and interfacial resistance to heat flow between the CNTs and polymer matrix.

In addition to thermal conductivity, the incorporation of CNTs in a polymer can also rectify the thermal stability (glass transition temperature, melting and thermal decomposition temperatures) as well as flame-retardant properties through their constraint effect on the polymer segments and chains. Again, these thermal properties are governed by many factors such as CNT type, aspect ratio and dispersion, its interfacial interactions with the polymer matrix, and so on. For example, adding 0.25 wt% silane-functionalized MWCNTs into epoxy resulted in a sharp rise in the glass transition temperature from 147 to 161°C, which was even higher than that of the nanocomposites containing CNTs without functionalization[13]. This capability becomes more pronounced when CNTs are incorporated into some thermoplastic matrix because CNTs can act as nucleation sites for crystallization of polymers. 1.0 wt% well-dispersed SWCNTs added into polymethyl methacrylate (PMMA) gave

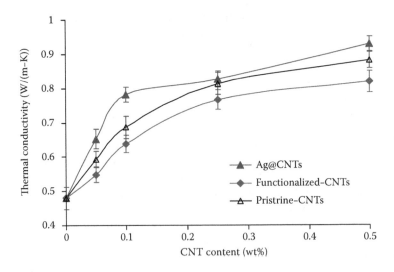

FIGURE 4.17
Comparison of thermal conductivities of composites containing CNTs with different functionalities: data in brackets indicate % increases in thermal conductivity against the polymer matrix. Adapted from Ma. P. C., et al. 2008. *Carbon* 46: 1497.

rise to a 40°C increase in glass transition temperature of PMMA[85], while the thermal decomposition temperature of polypropylene at peak weight loss was increased by 12°C with 2 vol% MWCNTs[86]. Kashiwagi et al.[87] studied the flammability of CNT/PMMA nanocomposites, showing that CNTs can outperform nanoclays as effective flame-retardant additives if they form a jammed network structure in the polymer matrix. The mechanisms behind this enhancement are twofold[87–89]: (1) the networks containing CNTs act as a heat shield for the neat polymer below the layer, thus significantly reducing the heat release rate of nanocomposites, and (2) the incorporation of CNTs in a polymer leads to an enhanced thermal stability of nanocomposites, thus effectively inhibiting the formation of cracks or openings that compromise the flame-retardant effectiveness of nanocomposites during burning. Poorly dispersed CNTs or very low concentration of CNTs resulted in the formation of a discontinuous layer consisting of fragmented islands with sizes from 1 to 10 mm[87]. The flame-retardant performance of the nanocomposites containing these island structures was much poorer than that of the nanocomposites with a continuous protective network layer. It is expected that the functionalization of CNTs will further reduce the flammability of nanocomposites as it will not only enhance the dispersion of CNTs in the polymer matrix, but also enhance the thermal stability of nanocomposites.

Thermomechanical properties refer to the mechanical responses of materials subjected to a temperature regime, and they can be evaluated using dynamic mechanical analysis (DMA). Ma et al.[13] studied the effects of CNT dispersion and functionalization on the thermomechanical properties of silane-functionalized CNT/epoxy, as shown in Figure 4.18. The addition of pristine CNTs to the polymer matrix showed little influence on storage modulus in the glassy region. In contrast, there was a stronger effect of silane-functionalized CNTs in the rubbery region at elevated temperatures where the improvements in elastic properties of nanocomposites were clearly observed (Figure 4.18A). This behavior can be explained in terms of interfacial interactions between the CNTs and epoxy. The improved interfacial interactions due to silane functionalization of CNT reduced the mobility of the local matrix material around the CNTs, increasing the thermal stability at elevated temperatures. There were marginal reductions in the storage modulus of the nanocomposites containing untreated CNT compared to the neat epoxy in the glassy region (Figure 4.18C), as a result of CNT agglomeration. In the rubbery region, however, the storage modulus increased with increasing CNT loading, similar to silane-CNTs.

The tan δ values obtained from the DMA analysis (Figures 4.18B and D) indicated that the peak tan δ for the neat epoxy was higher than that of the nanocomposites containing silane-CNTs, while it was comparable to that of the nanocomposites containing untreated CNT. Broad glass transition peaks and shoulders were noted from the tan δ curves of the nanocomposites containing silane-CNTs (curves B, C, and E in Figure 4.18B). These phenomena can be attributed to the fact that (1) the silane-treated CNT promoted the

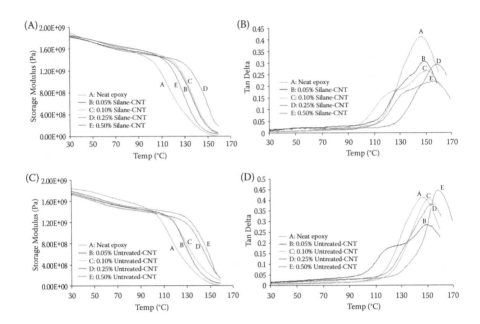

FIGURE 4.18
Thermodynamic mechanical properties of nanocomposites containing CNTs with different functionalities. Adapted from Ma. P. C., et al. 2007. *Compos. Sci. Technol.* 67: 2965.

cross-linking reactions of the epoxy and hardener, effectively discouraging the movement of molecular chains and (2) the covalent bonding between the silane-treated CNT and the epoxy promoted energy dissipation from the matrix to the CNT, resulting in an increase in loss modulus.

4.3.5 Optical Properties

The optical properties of CNT/polymer nanocomposites consist of optical nonlinearity and limitation, photoluminescence, light emission, and photovoltaic properties.

The nonlinear (NL) property of a material describes the behavior of light in nonlinear media in which the dielectric polarization responds nonlinearly to the electric field of light. Nonlinearity is typically observed at very high light intensities, such as those provided by pulsed lasers. It has been demonstrated that suspensions of both SWCNTs and MWCNTs have nonlinear optical properties[90,91], thus they can be used to rectify the optical properties of polymer composites to fabricate NL optical waveguide devices[97]. Tang et al.[93] pioneered this research field: wrapping of CNTs with poly(phenylacetylenes) (PPA) showed CNTs with a strong photostabilization effect and the PPA chains being protected from photodegradation under harsh laser irradiation with incident fluence. The CNT/PPA solutions effectively limited intense optical pulses, with the saturation fluence tunable by varying the CNT content. The NL properties of CNT/polymer nanocomposites are governed not only by the dispersion state of CNTs, but also by the polymer structure. Wang et al.[94] studied the NL properties of SWCNTs in N-methyl-2-pyrrolidone (NMP); N, N-dimethylformamide (DMF); and N, N-dimethylacetamide (DMA), and found that the NL extinction coefficients of SWCNTs markedly increased with increasing SWCNT concentration. CNTs dissolved in NMP exhibited a much better dispersion, although their optical limiting properties were inferior to those of DMF and DMA. The average bundle size of CNTs and physical properties of solvent dominated the NL extinction and optical limiting properties of CNT dispersions. Recent studies also revealed similar results: a good dispersion of CNTs did not guarantee a high performance of NL and optical limiting properties of CNT/polymer nanocomposites[95,96]. A possible reason for this observation is the complicated optical limiting mechanisms of composite materials. Besides the contribution by nonlinear scattering, other factors, such as nonlinear absorption and refraction and electronic absorption, also contributed to the optical limiting properties[97]. In addition to CNT dispersion, the backbones of polymers have a considerable effect on the NL properties of CNT/polymer nanocomposites[95]: for example, CNT functionalization using conjugated polymers having a high affinity with CNTs resulted in significant improvement in NL absorption and scattering due to the enhanced $\pi-\pi$ interactions between the CNTs and the polymer matrix.

The black color of CNTs combined with their high surface area and high electron affinity makes CNTs an ideal candidate to improve the photoluminescence of CNT/polymer nanocomposites. Photoluminescence is a process in which a substance absorbs photons, that is, electromagnetic radiation, and reradiates photons. In CNT/polymer nanocomposites, excitons are formed on the polymer by absorbing light and can be dissociated by electron transfer from the photoexcited polymer to CNTs. This process quenches the photoluminescence of polymers[98]. For example, Singh et al.[99] studied CNT/poly(3-hexylthiophene) nanocomposites, showing that the intensity of photoluminescence emission from the composite decreased with increasing CNT content and the absorption coefficient of the composites was insensitive to CNTs of up to 5 wt%. CNTs functionalized using phthalocyanines also showed a considerable quenching effect of the fluorescence intensity in the photoluminescence spectrum[100]. In contrast, earlier papers reported a strong luminescence from the MWCNTs functionalized with a variety of moieties. The reason for these inconsistencies may be attributed to the different defect states mediating luminescence by trapping excited electrons, which subsequently return to the valence band with the emission of light[101,102]. A recent study[98] revealed that visible photoluminescence from emissive impurities in nylon can be drastically enhanced when acid-treated MWCNTs are added (Figure 4.19). The origin of the luminescence was through the absorption by the acid-treated MWCNTs and energy transfer to emissive impurities intrinsic to the organic host matrix. Acid-treated MWCNTs were also shown to

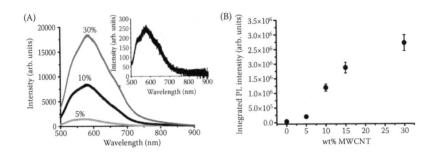

FIGURE 4.19
(A) Photoluminescence spectra of MWCNT/nylon composites for 488-nm excitation for different MWCNT contents. A broad luminescence band centered around 590 nm is observed (Inset: Photoluminescence spectra of a pure nylon film for 488-nm excitation); and (B) integrated PL intensity as a function of MWCNT content. Adapted from Henley, S. J., et al. 2007. *Small* 3: 1927.

enhance the resistance of organic chromophores to photobleaching. These results provide compelling evidence that MWCNTs have an enormous potential as a versatile material in future optoelectronic devices and can be considered a platform for both electronics and photonics.

Other potentially important optical applications of CNTs are polymer-based, light-emitting and photonic devices. The advantages of using organic light-emitting diodes (OLEDs) based on conjugated polymers include low cost, low operating voltage, excellent processability, and flexibility. However, their low quantum efficiency and low stability have limited wider applications and developments[103]. Kim et al.[104] observed that the device qualities, such as external quantum efficiency, improved by 2–3 times with the addition of up to 0.2 wt% SWCNTs in OLEDs. This observation was ascribed to the facile hole injection and the polymer–CNT interactions that stiffened the polymer chains. Subsequently, the better hole transport in the metallic SWCNT-polymer medium induced more efficient single exciton formation at or near the interface region. The single-layer LEDs fabricated using MWCNT-doped poly-p-phenylene benzobisoxazole decreased the threshold voltage by 2 V. The addition of 0.1 wt% of MWCNTs resulted in an increase in diode emission current by about two orders of magnitude over the LEDs without MWCNTs. However, a further increase of CNT content caused a successive decrease in light emission intensity due to the poor dispersion of CNTs[105].

Besides light-emitting applications, CNT/polymer nanocomposites are also widely used in photonic devices[103]. Arranz-Andrés et al.[106] investigated the effects of incorporating different types of CNTs, including single-walled, double-walled, and multiwalled CNTs, on the photovoltaic behavior of CNT/poly(3-hexiltiophene) nanocomposites. The results showed that the power conversion efficiency of the device made using these composites increased by three orders of magnitude compared with those without CNTs, but there was no clear correlation with the number of walls or diameter of CNTs. The highest open-circuit voltage and the best performance were achieved at a critical concentration of 5 wt% of CNT regardless of CNT type. Notice that the CNT content used in this study was about 5 wt%, which was much higher than the general CNT content of less than 1 wt% for polymer nanocomposites. It is expected that improved CNT dispersion can further enhance the photovoltaic properties of nanocomposites. In addition, charge transfer between the CNT and the polymer matrix plays an important role in converting photovoltaic properties of nanocomposite-based devices. Therefore, techniques for CNT functionalization using electron donors have also been developed recently, and some interesting results were generated by employing these functionalized CNTs[107–111].

Another interesting photonic application of CNT/polymer nanocomposites is the photomechanical actuator. The photomechanical effect is the change in the shape of a material when it is exposed to light[112]. Zhang et al.[113] demonstrated that SWCNT fibers can bend under light. This movement was sensitive to the intensity of the light, and was attributed to the localized electrostatic effects due to uneven distributions of photogenerated charges in CNTs. Actuators constructed using CNTs dispersed in several polymer matrices, such as Nafion, a sulfonated tetrafluoroethylene–based fluoropolymer-copolymer[114], and polydimethylsiloxane (PDMS)[115], have been reported. Figure 4.20A shows a schematic for the characterization of photomechanical actuation behavior of CNT/polymer nanocomposites. The introduction of CNTs in PDMS resulted in the composite actuator exhibiting better performance than those containing CB due to the high electrical conductivity of nanotubes (Figure 4.20B). Further studies revealed that the photomechanical response of CNT/polymer nanocomposites was an intrinsic behavior of CNTs, depending on the CNT alignment in the polymer matrix[115]. The incorporation of aligned CNT thin films in an acrylic elastomer allowed stable expansive actuation of 0.3% at a light illumination below 120 mW with an actuation stroke large enough to manipulate small objects[116]. Clearly, polymer/CNT nanocomposites represent a new class of multifunctional materials that are promising for organic photonic devices with improved performance.

4.3.6 Rheological Properties

Rheology is the study of the behavior of a material under conditions in which it flows rather than when it exhibits elastic or plastic deformation. It is also concerned with establishing predictions for mechanical behavior on a continuum mechanical scale based on the micro- or nanostructure of the materials[117]. The study of rheological response of CNT/polymer nanocomposites have both practical importance related to composite processing and scientific importance as a probe of the composite dynamics and microstructure[7].

The rheological properties of CNT/polymer nanocomposites depend on many factors such as filler content, aspect ratio and dispersion state, polymer molecular weight, and the interfacial interaction between the CNT

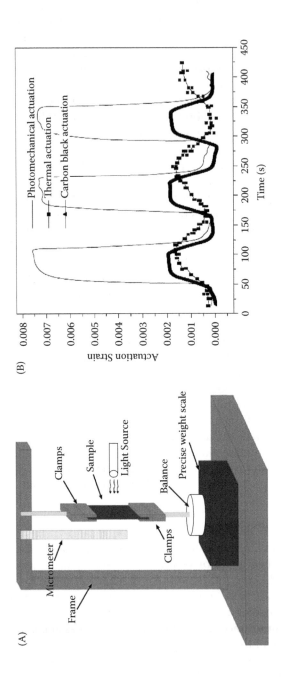

FIGURE 4.20

(A) Schematic of experimental setup for characterizing the photomechanical actuation, and (B) comparison of photomechanical actuation in SWCNT/PDMS multilayer actuators (solid line) with thermal actuation of the same structure (■) and the photomechanical actuation of a CB/PDMS multilayer actuator (▲). Adapted from Lu, S., et al. 2009. *J. Micro-Nano Mech.* 5: 29.

and the polymer[118–122]. The variations of viscosity and storage modulus of composites as a function of frequency are two characteristics that are commonly measured to present the rheological properties of CNT/polymer nanocomposites. Figure 4.21 shows the complex viscosity (Figure 4.21A) and storage modulus (Figure 4.21B) of CNT/PC nanocomposites with varying CNT contents[122]. At low frequencies, the fully relaxed polymer chains exhibit the typical Newtonian viscosity plateau. With the addition of CNTs, the low-frequency complex viscosity significantly increases, whereas the whole trend of viscosity drops with an increase of frequency at a given CNT content (Figure 4.21A), indicating that the relaxation of polymer chains in the nanocomposites is effectively restrained by the presence of CNTs. The storage modulus of nanocomposites gradually increases with increasing frequency and CNT content (Figure 4.21B), a reflection of a transition from viscous liquid to solid–like behavior. Similar to electrical percolation behavior, the rheology of a CNT/polymer nanocomposite also shows a transition from a rheological state (where the viscosity or storage modulus changes significantly with increasing filler content) to a solid–like behavior (where the viscosity or storage modulus is insensitive to or has only a slight variation with increasing filler content). The concentration at which this transition occurs is known as the *rheological percolation threshold*[119–122], which is shown in Figure 4.21C.

Different from the electrical percolation threshold, however, the rheological percolation threshold of CNT/polymer nanocomposites is strongly dependent on temperature. In MWCNT/PC nanocomposites, the percolation threshold decreased from 0.8 wt% to around 0.3 wt% upon increasing temperature from 210 to 300°C[122]. The reason for this observation is that the rheological response of nanocomposites is dominated by the networks formed by polymer chains and CNT–polymer interactions, and the mobility of these networks is significantly enhanced at elevated temperatures. In contrast, the electrical percolation threshold requires contacts of CNTs to form solid conducting networks where the temperature effect becomes marginal. Since the polymer chains contribute much to the rheological response of nanocomposites, it is easy to understand that the content of CNTs required to form a rheological percolation threshold should be much lower than that for electrical percolation. For example, Bose et al.[123] found that the rheological percolation threshold of MWCNT/PA6-ABS nanocomposites was around 1–2 wt%, which was significantly lower than the corresponding electrical percolation of around 3–4 wt%.

The rheological response of a nanocomposite containing a fixed CNT content below the percolation threshold can serve as an indirect measurement to show the degree of CNT dispersion in the polymer matrix. The better the dispersion of CNTs is, the lower the viscosity, in general. For example, Siddiqui et al.[124] successfully reduced the viscosity of MWCNT/epoxy suspension by functionalizing the CNTs using a nonionic surfactant, Triton-X100. The improved dispersion of CNTs in the epoxy matrix was mainly responsible

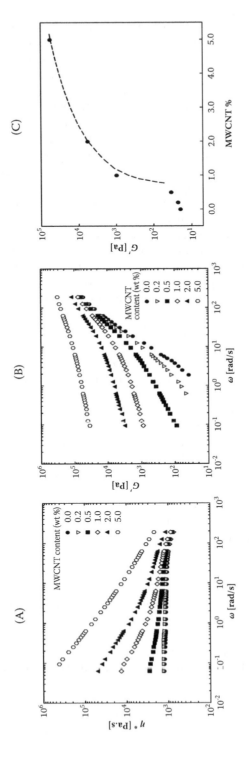

FIGURE 4.21
Rheological response of CNT/PC nanocomposites as a function of frequency at 230°C: (A) complex viscosity, (B) storage modulus, and (C) corresponding rheological percolation threshold at frequency of 1 rad/s. Adapted from Abbasi, S., et al. 2009. *Rheol. Acta.* 48: 943.

for the lowered viscosity. The surfactant consists of hydrophilic and hydrophobic segments: the hydrophobic octyl group of the surfactant interacts with CNTs through adsorption, while the hydrophilic segment links with epoxy through hydrogen bonding, effectively dispersing agglomerates into individual CNTs[55]. As such, functionalization of CNTs has the ameliorating effect of reducing the rheological percolation threshold of nanocomposites through improved CNT dispersion. Mitchell et al.[125] also showed improved SWCNT dispersion in PS by organic diazonium functionalization along with a concomitant drop in the rheological percolation threshold of CNT/PS composites from 3.0 to 1.5 wt%. The storage moduli at low frequencies were also higher for the nanocomposites containing functionalized CNTs, indicating enhanced load transfer between the CNT network and the polymer. The change in rheological properties of the polymer due to functionalization of CNTs indeed has a significant effect on reagglomeration behavior of CNTs in the polymer[35]. The pristine CNTs without functionalization started reagglomeration in the epoxy resin upon application of curing temperature, whereas the amino-functionalized CNTs remained uniformly dispersed over the whole curing process, indicating a beneficial effect of functionalization on the stability of CNT dispersion even at an elevated temperature. It is postulated that the amino-functional groups attached on the CNT surface affected the rheological behavior of epoxy, which in turn facilitated a faster cure and wrapping of a protective polymer layer on the CNT surface, preventing them from attracting one another.

4.3.7 Damping Properties

In contrast to the myriad research efforts to study the elastic properties of CNT/polymer nanocomposites, relatively less attention has so far been paid to their damping mechanisms and ability. *Damping* describes the trend of materials to reduce the amplitude of oscillations in an oscillatory system. As such, the damping properties of nanocomposites, such as loss factor and damping ratio, are essential design parameters for many engineering applications. High damping properties can be achieved in nanocomposites by taking advantage of the interfacial friction between the nanofillers and the polymer matrix[126–128]. The combination of extremely large surface area, weak interfacial bonding with the polymer, along with low mass density of the CNTs implies that frictional sliding of nanoscale tubes with the polymer matrix can cause significant dissipation of energy with a minimal weight penalty, providing a high damping capability of CNT-filled nanocomposites. Recent studies have shown that the damping properties of polymer nanocomposites were enhanced by more than 200% by incorporating a very small amount of CNTs[129–132]. The mechanism behind this enhancement can be explained by the *stick-slip theory*[126,133]: when the nanocomposite is subjected to an external load, shear stresses are generated between the CNT walls and the surrounding matrix due to the difference in their elastic properties according to the shear lag theory. CNTs are deformed together with the matrix if they are well bonded. However, when the external load exceeds a critical value, leading to debonding of CNTs from the matrix, the CNTs will stop elongating together with the matrix and a further increase of the load can only result in the deformation of the matrix. Thus, the polymer starts to flow over the surface of CNTs and deformation energy will be dissipated through the slippage between the CNTs and the matrix. This phenomenon leads to the enhanced damping properties of nanocomposites compared to the neat polymer.

There are several factors governing the damping properties of CNT/polymer nanocomposites. Suhr et al.[134] studied the damping properties of PC-based nanocomposites filled with SWCNTs and MWCNTs. While both of the CNTs gave rise to higher damping than the neat PC, the loss modulus of the nanocomposite containing SWCNTs was significantly higher than that filled with MWCNTs, suggesting that the inner layers of MWCNTs did not contribute to interfacial frictional sliding with the matrix for energy dissipation. The effect of CNT dispersion on the damping properties of nanocomposites shows a trend similar to their elastic mechanical properties: improved CNT dispersion can enhance the damping properties of nanocomposites. As shown in Figure 4.22A, sonication for a longer time of 15 minutes resulted in a better CNT dispersion than sonication for 2 min, and the corresponding composites exhibited much enhanced loss modulus compared to the counterpart with poor CNT dispersion under the same testing condition[131].

As described above, a weak interfacial adhesion plays an important role in enhancing the damping properties of CNT nanocomposites, although it is detrimental to the mechanical properties of CNT/polymer nanocomposites. When the interfacial bond between the pristine CNTs and epoxy was weak with its shear strength of about 0.5 MPa, up to a 1400% increase in damping ratio was obtained for the nanocomposites compared to the neat polymer[129].

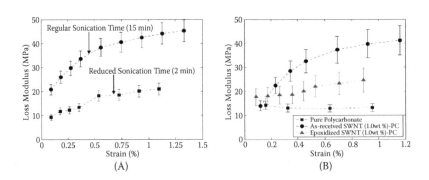

FIGURE 4.22
(A) Effect of CNT dispersion and (B) functionalization on the damping properties of CNT/PC nanocomposites. Adapted from Suhr, J., et al. 2008. *J. Mater. Sci.* 43: 4370.

In contrast, a stronger interfacial bond arising from functionalization inhibited filler-matrix friction, thereby lowering the damping response, as shown in Figure 4.22B[131]. In other words, covalent bonding of CNTs with the polymer matrix prevented the activation of interfacial sliding leading to enhanced storage modulus and reduced loss modulus.

As discussed in Chapter 2, CNT alignment can augment the strength and modulus of CNT/polymer nanocomposites. However, it may not be beneficial to the enhancement of damping properties of nanocomposites because randomly dispersed CNTs can be subject to interfacial shear stresses from all directions due to the same load from the initial vibration source[135,136]. Thus, it is expected that the higher load applied to the nanocomposite containing random-dispersed CNTs, the more CNT–matrix interfacial slippage is to be initiated, and thus the higher the energy dissipation will be.

4.4 CNT/Polymer Interface

4.4.1 Importance of the Interface

An *interface* is a surface forming a common boundary between two different phases, such as an insoluble solid and a liquid, two immiscible liquids, or a liquid and an insoluble gas. The property of the interface depends on the type and dimension of the system being treated: the bigger the quotient area/volume, the more effect the surface phenomena will have. The interface has to be considered in systems consisting of fillers with a big surface area-to-volume ratio, such as CNT/polymer nanocomposites. Figure 4.23 shows schematically the interfacial regions as a function of filler size: large particles produce a low radius of curvature and relatively less polymer in the interfacial region, and under the same volume of filler broken into smaller particles, a higher radius of curvature is created and more polymer is involved in the interfacial region (Figure 4.23B)[137]. The interface is a region with altered chemistry, altered polymer chain mobility, altered degree of cure, and altered crystallinity that are unique compared to those of the filler or the matrix. It has been clearly demonstrated in Chapter 2 that the nanoscopic dimensions of CNT particles and their large quantities for a given volume fraction lead to an exceptionally large surface area and interface region in a composite. This in turn means that the design and control of the interface is particularly important for CNT/polymer nanocomposites as in many fiber reinforced composites[138].

As discussed in Chapters 2 and 3, CNTs possess excellent mechanical properties, but these properties alone do not guarantee mechanically superior nanocomposites because the mechanical characteristics of nanocomposites depend not only on the properties of reinforcements, but more importantly on the degree to which an applied load is transferred from the matrix phase

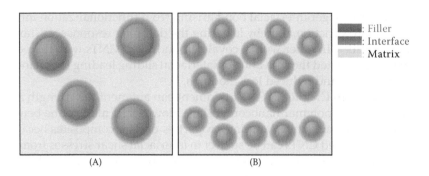

**: Filler
: Interface
: Matrix

FIGURE 4.23
Interfacial regions as a function of filler particle size: (A) relatively less polymer in the interfacial region of large particles, and (B) more polymer is involved in the interfacial region of particles under the same volume of filler broken into smaller particles. Adapted from Schadler, L. 2007. *Nature Mater.* 6: 257.

to the reinforcements through the interface, that is, the mechanics that govern the interfacial region between the CNT and the polymer[139–143]. Take the transverse crack propagation as an example: when the crack reaches the interface, it tends to propagate along the interface because the interface is relatively weaker than the filler (with respect to resistance to crack propagation). If the interface is weaker than the matrix, the crack will cause the interface to fracture and result in failure of the composite. In this aspect, CNTs are better than traditional fibers, such as glass and carbon fibers, due to their ability to inhibit nano- and microscopic cracks. Hence, knowledge and understanding of the nature and mechanics of the interface between the CNTs and polymer are critical to predicting and manufacturing mechanically enhanced CNT/polymer nanocomposites. This will also enable material scientists and engineers to tailor the interface for specific end applications or superior mechanical properties.

4.4.2 Methodologies for Studying the CNT–Polymer Interface

There are three major modes of interactions between a CNT and a polymer matrix: covalent bonding, nano-mechanical interlocking, and noncovalent interactions such as van der Waals and electrostatic forces. The perfect carbon–carbon bond along with the inherently inert nature of carbon atoms limits any sort of strong covalent bonding of CNTs with a surrounding polymer matrix. The interactions of CNTs with a given polymer matrix can be improved via chemical modification of the CNT surface with functional groups. However, a disadvantage of modifying the hollow nanotubes by covalent functionalization is that it changes the surface structure and breaks carbon–carbon bonds along the sidewall, and therefore affects the intrinsic properties of the nanotubes. Fortunately, noncovalent functionalization of CNTs such as wrapping of surfactants and polymers around the CNTs or adsorption of aromatic structures onto their sidewalls via π-π stacking can often be adopted to improve the level of interaction at the interfaces. Nano-mechanical interlocking could be difficult in CNT/polymer nano-composites due to the atomically smooth surface of CNTs. Some evidence has proven that nanotubes slide relatively easy from a surrounding polymer matrix after full debonding from the matrix[144]. However, this might not be the case with nanotubes with twisted, uneven surfaces or mechanically deformed nanotubes with surface steps formed by gliding of dislocations. Some form of interlocking may exist in these systems.

The determination of the CNT–polymer interface experimentally is a challenging task because of the technical difficulties associated with the manipulation of nanoscale objects. Broadly based studies on the interfacial mechanics of CNT/polymer nanocomposites are appealing from three aspects: mechanics, chemistry, and physics[139–143].

From a mechanics point of view, the important questions are as follows:

1. Is there evidence of stress transfer and what are the expected values? What experimental techniques can be developed to measure the interfacial bond strength and adhesion on the nanoscale?
2. The relationship between the mechanical properties of individual constituents, that is, the CNT and polymer, and the properties of the interface and the bulk composite.
3. The effect of the unique length scale and structure of CNTs on the property and behavior of the interface.
4. The ability of the mechanics modeling to estimate the properties of the composites for the design process for structural applications.

From a chemistry point of view, the interesting issues are as follows:

1. Is there evidence of adequate wetting of CNTs by polymers? Is this wetting enough to offer a sufficient condition for good adhesion between the CNTs and the polymer matrix? If yes, then what is the nature of this wetting, hydrophobic or hydrophilic interactions?
2. The chemistry of the bonding between the CNTs and the polymer matrix, especially the nature of bonding, for example, covalent or noncovalent interaction.
3. The relationship between the processing and fabrication conditions of the CNT/polymer nanocomposite and the resulting chemistry of the interface.
4. The effect of CNT functionalization on the mechanical properties of CNTs and the corresponding nature and strength of the bonding at the CNT–polymer interface.

From the physics point of view, researchers are interested in the following issues:

1. The CNT–polymer interface serves as a model nano-mechanical or a lower dimensional system (one dimension). It would be interesting to understand the nature of the dominating forces at the nanoscale and the effect of surface forces that are expected to be significant due to the large surface-to-volume ratio of CNTs.

2. The length scale effects on the interface and the differences between the phenomena of mechanics at the macro- (or meso-) and the nanoscale.

3. What are the possible molecular mechanisms for the CNT–polymer interface? Is there any correlation between the interfacial bond strength and the CNT geometry, such as wall thickness and chirality, aspect ratio, and the properties of the external graphene surface?

Computational techniques have extensively been used to study the interfacial mechanics and nature of bonding in CNT/polymer nanocomposites. The computational studies can be broadly classified as atomistic simulations and continuum methods. The atomistic simulations are primarily based on molecular dynamic (MD) simulations and density functional theory (DFT)[140,143]. The main focus of these techniques is to understand and study the effect of bonding between the CNT and polymer, that is, covalent, electrostatic, or van der Waals forces, and the effect of friction on the CNT–polymer interface. Liao et al.[145] studied the interfacial characteristics of CNT/polystyrene (PS) nanocomposites using MD simulations and elasticity calculations. They found that in the absence of atomic bonding between a CNT and a polymer matrix, electrostatic and van der Waals interaction, deformation induced by these forces, and stress and deformation arising from a mismatch in the coefficients of thermal expansion, were the three major noncovalent interactions and bonding mechanisms. All of these interactions contribute to the ability of interfacial stress transfer, and they can be regarded as the critical parameters in controlling the mechanical performance of nanocomposites. The results from a CNT pull-out simulation also showed that the interfacial strength of the CNT-PS system was about 160 MPa, significantly higher than most carbon fiber–reinforced polymer composite systems. This result was comparable to the CNT–polymer interfacial strength by covalent bonding, possibly because the authors did not notice the π–π interaction between the CNTs and the aromatic rings in PS.

It should be noted there were large variations on the principles and assumptions of MD simulations; consequently some conflicting results regarding the effect of noncovalent bonding on the interfacial strength can be found in literature. For example, Namilae et al.[146] presented an algorithm for parallel implementation of MD simulations of CNT-based systems to model the CNT–polymer interface with three levels of interactions, including long-range van der Waals interactions, chemically bonded with a fixed matrix and chemically bonded with a matrix. It is shown that the interfacial bond strength arising from the noncovalent interactions was very low, on the order of few MPa, and it was significantly improved through chemical functionalization of CNTs, to an order of a few GPa. It is further noted that chemical bonding between the functionalized CNTs and the matrix during processing was essential to guarantee strong interfacial bonds and hence achieve composites with excellent mechanical properties. Liu et al.[147] introduced a hybrid system integrating both covalent and noncovalent functionalization for interfacial design of CNT/polymer nanocomposites. To investigate the feasibility of this system, epoxidized SWCNTs followed by wrapping with poly (*m*-phenylenevinylene-alt-*p*-phenylenevinylene) (PmPV) were studied by MD simulation, and the major findings are summarized in Table 4.4. It is shown that PmPV molecules were miscible with epoxy resin and tended to wrap around the epoxidized SWCNTs, which in turn encouraged CNT dispersion as well as its interaction with the matrix. The interfacial bond strength of CNT/polymer nanocomposites surged to 940 MPa after hybrid functionalization using epoxy and PmPV, which is the highest among the four different cases of surface condition, suggesting that it is indeed possible to improve the CNT–polymer interfacial strength by a molecularly designed functionalization process.

The continuum methods extend the continuum theories of micromechanics modeling and fiber-reinforced composites to CNT/polymer nanocomposites and explain the behavior of the composite from a mechanics point of view. Currently, these approaches seem to be the only feasible method for large-scale analysis[140–143]. Chen et al.[148] studied the mechanical properties of CNT-based nanocomposites using modified continuum mechanics and the finite element method (FEM). Numerical results using the FEM showed that the load-carrying capacities of the CNTs in a matrix were significant. For example, with the addition of CNTs in a matrix at a volume fraction of 3.6%, the stiffness of the composite increased as much 33% in the axial direction with long CNTs. These simulation results are consistent with the experimental results reported in the literature and the results obtained from a modified continuum mechanics. In order to make an accurate simulation using a continuum mechanics model, many factors, such as the dimension and aspect ratio of CNTs and the number of walls in CNTs should be taken into account. Wagner[149] used a modified continuum model taking into consideration the

TABLE 4.4 MD Simulation Results on the Interfacial Strength of SWCNTs/Epoxy Nanocomposites.

SWCNT	Interfacial strength (MPa)
Pristine SWCNT	170
PmPV wrapped SWCNTs (Non-covalent functionalization)	290
Epon 828 epoxidized SWCNT (Covalent functionalization)	690
Hybridized functionalized SWCNT (Covalent + Noncovalent functionalization)	940

Source: Liu, J. Q., et al. 2007. *Nanotechnology* 18: 165701.

tube structure of SWCNTs to study the CNT–polymer interface, and found that a high interfacial bond strength was obtained in SWCNT-reinforced nanocomposites, which was in agreement with the results from MD simulations. However, MWCNT-reinforced nanocomposites were not considered because of the possible telescopic interwall sliding failure that occurred in MWCNTs even at low strength of 0.5 MPa. It was noted that although the simulation models shed considerable light on the CNT–matrix interface, the question still remains as to how to accurately capture the physical essence of the filler–matrix interface[143].

On the experimental front, the evaluation of the CNT–polymer interface can be accomplished by employing such techniques as Raman spectroscopy and a CNT pull-out experiment using atomic force microscopy (AFM) or scanning probe microscopy (SPM). In the field of fiber-reinforced polymers, it is well known that the application of a mechanical strain to carbon or Kevlar fiber results in shifted wavenumbers of the Raman peaks, which are directly related to the interfacial interactions between the fiber and matrix[138]. Correlating such shifts with the applied strain (through a calibration procedure) leads to the determination of interfacial strength profiles in the embedded fibers. Similar Raman shifts have also been observed when CNT/polymer nanocomposites were loaded in tension[150–155]. As discussed in Chapter 1, the Raman spectroscopy is a commonly used technique to characterize the sp^2 structure of CNTs. The G′ band, the first overtone of the defect-induced D band observed in the range of 2450–2750 cm^{-1}, can be related to the mechanical deformation of nanotubes, thus offering a powerful tool to investigate the interfacial properties between the CNTs and polymer matrix.

The tensile and compressive properties of MWCNT/epoxy nanocomposites were measured based on the Raman spectra[150]. The G′ band peak of nanocomposites was shifted upward by 7 cm^{-1} under a compressive strain, whereas a slightly positive shift was obtained in tension. Figure 4.24 summarizes these G′ band shifts, as a reflection of the strain transferred from the matrix to the CNTs. The different Raman responses in tension and compression likely arose from the peculiar structure of MWCNTs: in tension, the outer layer of the MWCNT is loaded, but the load is not effectively transferred to the inner layers due to the relatively weak bonds between the nanotube layers and slippage between the inner and outer tubes. (The slippage has been confirmed by the CNT pull-out experiment, as discussed in the following.) Under compression, however, the load transfer from the outer layer to the inner layers of the MWCNT is possible through the buckling and bending of nanotubes, and the slippage between the nanotube layers is discouraged because of the seamless structure of the tubes and the geometrical constraint imposed on the inner layers by the outer graphene layer. Similar experiments were also performed on SWCNTs and double-walled CNTs (DWCNTs), but the Raman results showed that there was almost no shift of the G′ band under compression and a significant downward shift in tension (Figure 4.25)[155]. The investigation of the G′ band of the inner wall of DWCNTs showed that the load

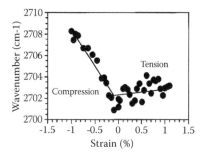

FIGURE 4.24

Raman wavenumber of MWCNT/epoxy nanocomposites as a function of applied strain in tension and compression. Adapted from Schadler, L. S., et al. 1998. *Appl. Phys. Lett.* 73: 3842.

transfer from the outer layer to the inner wall of DWCNTs in nanocomposites was very poor, and the inner wall was virtually unstressed during both tensile and compressive deformation (Figure 4.25A)[155]. The implication is that only the outermost wall of MWCNTs contributes to the load transfer between the CNTs and the polymer matrix.

However, there were some inconsistencies regarding experiment observations and the explanation of the observed G′ band shift of CNT-filled nanocomposites. It was reported[156] that the applied compressive load transferred into CNTs through buckling, bending, or twisting of the nanotubes is sensitive to Raman spectroscopy. Because as-received SWCNTs are mostly in the form of ropes or agglomerates due to poor dispersion, the small Raman shift obtained in tension only reflects slippage of individual SWCNTs within the agglomerates. For the nanocomposites containing random dispersed CNTs, there is a possibility that nanotubes lying in the tensile direction are loaded in tension, but those in the perpendicular direction are under compression because of the Poisson's contraction[154]. In addition, the laser spot size used for Raman spectroscopy is about 1–5 μm, which is much larger than the diameters of CNTs; thus the Raman signal is averaged over nanotubes present in all directions. This implies that all of these different situations should be considered when explaining the results obtained from Raman spectroscopy.

In summary, an empirical linear relationship exists between the G′ band shift and the applied elastic strain of CNT/polymer nanocomposites[154–157], as expressed in the following equation:

$$\Delta N_w = m\varepsilon \tag{4.2}$$

where ΔN_w is the nanotube G′ band shift between zero strain and the applied strain ε (ΔG′ in Figure 4.25), and m is the corresponding slope. This slope is regarded as an indication of the efficiency of interfacial stress transfer. The higher the absolute value of m, the better the efficiency of load transfer from

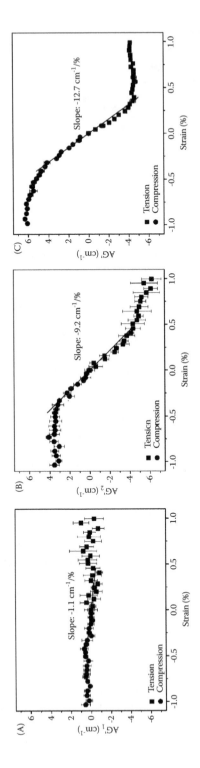

FIGURE 4.25
Shift of Raman G′ bands as a function of tensile and compressive strains applied to different CNT/epoxy matrix composites: (A) inner walls of DWCNTs, (B) outer walls of DWCNTs, and (C) SWCNTs (ΔG′ refers to the band shift between zero strain located at 2630 cm⁻¹ and the applied strain). Adapted from Cui, S., et al. 2009. *Adv. Mater.* 21: 3591.

the matrix to the CNTs. When incorporating the mechanical stress-strain behavior of nanocomposites (Eq. [4.3]) with the Raman measurement, it can be further deduced that the strength of nanocomposites can be expressed as follows:

$$\sigma = E\varepsilon \tag{4.3}$$

$$\sigma = (E\Delta N_w)/m \tag{4.4}$$

where E is the Young's modulus of the nanocomposites. Therefore, the Raman spectrometry offers a qualitative means to evaluate the effect of load transfer on the strength of the nanocomposites. The importance of this correlation lies in the Raman spectroscopy being able to offer a way to detect the elastic stress or strain in the nanocomposites. For example, it may be useful to measure the matrix stress distribution in the vicinity of fibers (by coating fibers with CNTs) to detect or predict the onset of failure in composite materials[158], and thus provide a means to estimate the interfacial shear strength between the CNTs and the polymer matrix.

In situ straining measurements have also been used to understand the mechanics, fracture, and failure processes of the CNT–polymer interface. In these techniques, the CNT/polymer nanocomposite is strained by an SPM or AFM tip inside a TEM or SEM chamber and simultaneously imaged to get real-time and spatially resolved information[144,159–162]. Figure 4.26 shows an optical micrograph and the principle of this measurement system inside a TEM[163]. In the specimen area of the holder, the right-hand side is the mobile part and the left-hand side is the fixed part. The mobile part is connected to a tube-type piezoelectric device for measuring displacements on the atomistic scale in three directions (i.e., x, y, and z directions), and a microscrew motor for measuring coarse displacements. To perform the tensile test inside a TEM machine, an aluminum plate is mounted between the mobile and fixed components using a silver paste for easy handling.

Cooper et al.[144] used a technique similar to the one previously described to measure the interfacial strength, which involved drawing out individual MWCNTs and SWCNT ropes that bridged the propagating crack in the matrix (Figure 4.27B). Tests on MWCNT/epoxy nanocomposites showed that the measured interfacial strength decreased with an increasing embedded length of nanotube in the matrix (Figure 4.27A), which was significantly higher than those reported for fiber–polymer interfaces (9–64 MPa)[164]. In some cases, the interfacial strength was as high as 376 MPa, resulting in the pull-out of inner walls of MWCNTs (Figure 4.27C). Meanwhile, only one SWCNT rope specimen underwent pull-out while all other samples resulted in SWCNT fracture, indicating that the interfacial bond strength of SWCNT/ epoxy exceeded the SWCNT rope strength. These observations on both the interfacial strength for MWCNTs and the breaking strength of SWCNTs

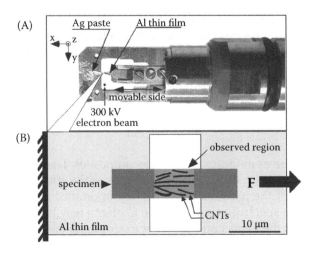

FIGURE 4.26
(A) Optical micrographs of the specimen holder for in situ strain measurements on a TEM, and (B) the illustration of the principle of the tensile system inside the TEM chamber. Adapted from Deng, F., et al. 2007. *Compos. Sci. Technol.* 67: 2959.

ropes suggested excellent load transfer characteristics in CNT/polymer nanocomposites.

The aforementioned technique was modified by Barber et al.[161] to perform a single-fiber pull-out test to measure the interfacial strength of MWCNT/polyethylene-butene nanocomposites. The individual CNTs attached to the end of an AFM tip were pushed into a molten polymer. After solidifying the polymer matrix, the nanotubes were pulled while monitoring the pull-out force. Using an energy balance approach, the interfacial fracture energies of the pristine nanotubes were found in the range of 4–70 J/m², which were comparable to that of glass fiber pull-out from polymers (6–61 J/m²)[165]. These interfacial strength values were considered to be high due to the bonding of CNTs with the polymer via defects in the tube wall or the wrapping of polymer chains around the nanotubes. Similar pull-out experiments were also carried out to study the effect of chemical functionalization of CNTs on the interfacial bond strength of CNT/epoxy nanocomposites[162]. The results showed a higher pull-out force and a shorter critical embedded length for fracture of CNTs when carboxyl-functionalized CNTs were tested than those without. These observations clearly indicated that CNT functionalization was an effective way of enhancing the interfacial bond strength between CNTs and the polymer matrix, and the stress transfer was more efficient in CNT/polymer nanocomposites than those observed in conventional microscale fiber-based composites.

While it has been nearly a decade since the first measurement was made on the interfacial strength of the CNT–polymer interface, and many studies have been devoted to understanding the load-transfer mechanisms in

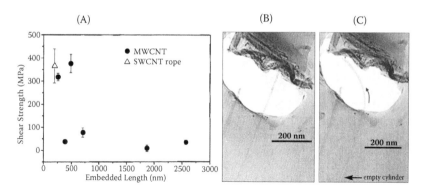

FIGURE 4.27
Direct measurements of interfacial bond strengths between CNTs and an epoxy matrix: (A) correlation between interfacial shear strength and embedded CNT length in nanocomposites, (B) TEM image of a nanotube bridging the matrix hole, and (C) TEM image of the same specimen following partial pull-out by means of an SPM tip. (The curved arrow shows the direction of the tip movement, and the straight arrow indicates the empty cylindrical hole left behind after partial pull-out.) Adapted from Cooper, C. A., et al. 2002. *Appl. Phys. Lett.* 81: 3873.

CNT/polymer nanocomposites, relatively little progress has been achieved. The ability to manipulate the behavior of the CNT/polymer matrix interface to achieve desired properties still remains rather limited and largely empirical[143]. The difficulties originate from two fundamental issues: (1) reliable measurements of the critical interfacial parameters, such as shear bond strength, are difficult to make, given the small size of CNTs and the lack of a robust testing platform on the nanoscale and (2) establishment of the theoretical framework connecting nanoscale interfacial features to macroscopic properties has not been fully developed. Obviously, the design and control of the CNT–polymer interface are two of the most challenging areas in the practical application of CNT/polymer nanocomposites.

References

1 Ajayan. P. M., et al. 2003. *Nanocomposite science and technology*, 77–80. New York: Wiley-VCH.
2 Santos. A. S., et al. 2008. *J. Appl. Polym. Sci.* 108: 979
3 Ajayan. P. M., et al. 1994. *Science* 265: 1212.
4 Ma, P.C. et al. 2010. Compos A 41: 1345.
5 Du. J. H., et al. 2007. *eXPRESS Polym. Lett.* 1: 253.
6 Grossiord. N., et al. 2006. *Chem. Mater.* 18: 1089.
7 Moniruzzaman. M., et al. 2006. *Macromolecules* 39: 5194.
8 Zhang. Q. H., et al. *J. Mater. Sci.* 39: 1751.
9 Hill. D. E., et al. 2002. *Macromolecules* 35: 9466.

10 Kim. J. Y., et al. 2006. *J. Polym. Sci.* B 44: 1062.
11 Gojny. F. H., et al. 2004. *Compos. Sci. Technol.* 64: 2303.
12 Moisala. A., et al. 2006. *Compos. Sci. Technol.* 66: 1285.
13 Ma. P. C., et al. 2007. *Compos. Sci. Technol.* 67: 2965.
14 Ma. P. C., et al. 2008. *Carbon* 46: 1497.
15 Kosmidou. T. V., et al. 2008. *eXPRESS Polym. Lett.* 2: 364.
16 Ma. P. C., et al. 2009. *ACS Appl. Mater. Interfaces* 1: 1090.
17 Jiang. K. L., et al. 2002. *Nature* 419: 801.
18 Zhang. X. B., et al. 2006. *Adv. Mater.* 18: 1505.
19 Liu. K., et al. 2008. *Nano. Lett.* 8: 700.
20 Chen. Q. F., et al. 2010. *Carbon* 48: 260.
21 Wardle. B. L., et al. 2008. *Adv. Mater.* 20: 2707.
22 Vigolo. B., et al. 2000. *Science* 290: 1331.
23 Mamedov. A. A., et al. 2002. *Nature. Mater.* 1: 190.
24 Xia, H., et al. 2004. J. Appl. Polym. Sci. 93: 378.
25 Masuda, J., et al. 2008. *Macromolecules* 41: 5974.
26 Bryning, M. B., et al. 2005. *Adv. Mater.* 17: 1186.
27 Li, J., et al. 2007. *Adv. Funct. Mater.* 17: 3207.
28 Ramanathan, T., et al. 2005. *Chem. Mater.* 17: 1290.
29 Wang, S., et al. 2006. *Nanotechnology* 17: 1551.
30 Sham, M. L., et al. 2006. *Carbon* 44: 768.
31 Ma, C., et al. 2008. *Carbon* 46: 706.
32 Zhu, Y. F., et al. 2009. *J. Appl. Phys.* 105: 054319-1.
33 Shofner, M. L., et al. 2006. *Chem. Mater.* 18: 906.
34 Chen, G. X., et al. 2008. *Polymer* 49: 943.
35 Ma, P. C., et al. 2010. *Carbon* 48: 1824.
36 May, C. A. 1988. *Epoxy resins: Chemistry and technology* (2nd ed), 907. New York: Marcel Dekker.
37 Abdalla, M., et al. 2008. *Polymer* 49: 3310.
38 Coleman, J. N., et al. 2006. *Adv. Mater.* 18: 689.
39 Qian, D., et al. 2000. *Appl. Phys. Lett.* 76: 2868.
40 Jonathan, N., et al. 2006. *Carbon* 44: 1624.
41 Jiang, B., et al.20007. *Compos. B* 38: 24.
42 Hernández-Pérez, A., et al. 2008. *Compos. Sci. Technol.* 68: 1422.
43 Liu, L. Q., et al. 2005. *Compos. Sci. Technol.* 65: 1861.
44 Ma, P. C., et al. 2008. Effect of dispersion techniques and functionalization of CNTs on the properties of CNT/epoxy composites. Paper presented at the 2nd International Conference on Advanced Materials and Structures (ICAMS-2), Nanjing, China, October 2008.
45 Thostenson, E. T., et al. 2002. *J. Phys. D* 35: 77.
46 Haggenmueller, R., et al. 2003. *J. Nanosic. Nanotechnol.* 3: 105.
47 Meng, H., et al. 2008. *Polymer* 49: 610.
48 Yang, B. X., et al. 2009. *J. Appl. Polym. Sci.* 113: 1165.
49 Yang, B. X., et al. 2007. *Adv. Funct. Mater.* 17: 2062.
50 Yuen, S. M., et al. 2007. *Compos. Sci. Technol.* 67: 2564.
51 Koval'chuk, A. A., et al. 2008. *Macromolecules* 41: 7536.
52 Byrne, M. T., et al. 2008. *Nanotechnology* 19: 415707.
53 Paiva, M. C., et al. 2004. *Carbon* 42: 2849.
54 Kim, K. H., et al. 2008. *Compos. Sci. Technol.* 68: 2120.

55 Geng, Y., et al. 2008. *Compos. A* 39: 1876.
56 Xia, H., et al. 2006. *J. Mater. Chem.* 16: 1843.
57 Sahoo, N. G., et al. 2006. *Macromol. Chem. Phys.* 207: 1773.
58 Sui, G., et al. 2008. *Mater. Sci. Engin. A* 485: 524.
59 Shanmugharaj, A. M., et al. 2007. *Compos. Sci. Technol.* 67: 1813.
60 Breuer, O., et al. 2004. *Polym. Compos.* 25: 630.
61 Ping, Z. Y., et al. 2001. *Polymer world*, 183. Shanghai: Fudan University Press.
62 Zhang, W., et al. 2007. *J. Mater. Sci.* 42: 3408.
63 Aneli, J. N., et al. 1998. *Structuring and conductivity of polymer composites*, 50. Commack, NY: Nova Science.
64 Li, J. 2006. Electrical conducting polymer nanocomposites containing graphite nanoplatelets and carbon nanotubes. PhD thesis. Hong Kong University of Science and Technology 1–100.
65 Ma, P. C. 2008. Novel surface treatment, functionalization and hybridization of carbon nanotubes and their polymer-based composites. PhD thesis. Hong Kong University of Science and Technology 1–30.
66 Gul, V. E. 1996. *Structure and properties of conducting polymer composites*, 27–123. VSP.
67 Bauhofer, W., et al. 2009. *Compos. Sci. Technol.* 69: 1486.
68 Sandler, J. K. W., et al. 2003. *Polymer* 44: 5893.
69 Martina, C. A., et al. 2005. *Polymer* 46: 877.
70 Buldum, A., et al. 2001. *Phys. Rev. B* 63: 161403.
71 Stadermann, M., et al. 2004. *Phys. Rev. B* 69: 201402.
72 Li, J., et al. 2008. *Mater. Sci. Engin. A* 483-484: 660.
73 Peng, H., et al. 2008. *Small* 4: 1964.
74 Bokobza, L., et al. 2008. *J. Polym. Sci. B* 46: 1939.
75 Fan, Z. J., et al. 2008. *New Carbon Mater.* 23: 149.
76 Sumfleth, J., et al. 2009. *J. Mater. Sci.* 44: 3241.
77 Sun, Y., et al. 2009. *Macromolecules* 42: 459.
78 Phillpot, SR, et al. 2005. *Mater. Today* 8: 18.
79 Biercuk, M. J., et al. 2002. *Appl. Phys. Lett.* 80: 2767.
80 Evseeva, L. E., et al. 2008. *Mech. Compos. Mater.* 44: 87.
81 Guthy, C., et al. 2007. *J. Heat Trans.* 129: 1096.
82 Shenogin, S., et al. 2004. *J. Appl. Phys.* 95: 8136.
83 Cai, D., et al. 2008. *Carbon* 46: 2107.
84 Gojny, F. H., et al. 2006. *Polymer* 47: 2036.
85 Velasco-Santos, C., et al. 2003. *Chem. Mater.* 15: 4470.
86 Kashiwagi, T., et al. 2002. *Macromol. Rapid Commun.* 232: 761.
87 Kashiwagi, T., et al. 2005. *Nature Mater.* 4: 928.
88 Kashiwagi, T., et al. 2005. *Polymer* 46: 471.
89 Schartel, B., et al. 2005. *Eur. Polym. J.* 41: 1061.
90 Riggs, J. E., et al. 2000. *J. Phys. Chem. B* 104: 7071.
91 O'Flaherty, S. M., et al. 2003. *J. Opt. Soc. Am. B* 20: 49.
92 Sakakibara, Y., et al. 2006. Carbon nanotube polymer nanocomposite and its nonlinear optical applications. European Conference on Optical Communications Proceedings, Paper ID: 4801138.
93 Tang, B. Z., et al. 1999. *Macromolecules* 32: 2569.
94 Wang, J., et al. 2008. *J. Phys. Chem. C* 112: 2298.
95 Zhang, L., et al. 2009. *J. Phys. Chem. C* 113: 13979.

96   Qiu, X. Q., et al. 2008. *Chinese Phys. Lett.* 25: 536.
97   Chen, Y., et al. 2007. *J. Nanosci. Nanotechnol.* 7: 1268.
98   Henley, S. J., et al. 2007. *Small* 3: 1927.
99   Singh, I., et al. 2008. *Carbon* 46: 1141.
100 He, N., et al. 2009. *J. Phys. Chem. C* 113: 13029.
101 Lin, Y., et al. 2005. *J. Phys. Chem. A* 109: 14779.
102 Guldi, D. M., et al. 2003. *Chem. Commun.* 10: 1130.
103 Baibarac, M., et al. 2006. *J. Nanosci. Nanotechnol.* 6: 1.
104 Kim, J. Y., et al. 2003. *Opt. Mater.* 21: 147.
105 Huang, J. W., et al. 2005. *Nanotechnology* 16: 1406.
106 Arranz-Andrés, J., et al. 2008. *Carbon* 46: 2067.
107 Valentini, L., et al. 2005. *J. Appl. Phys.* 97: 114320.
108 Valentini, L., et al. 2006. *J. Appl. Phys.* 99: 114305.
109 Xu, W. J., et al. 2006. *Nanotechnology* 17: 728.
110 Yun, D., et al. 2008. *Synth. Met.* 158: 977.
111 Kalita, G., et al. 2009. *Current Appl. Phys.* 9: 346.
112 Lu, S., et al. 2009. *J. Micro-Nano Mech.* 5: 29.
113 Zhang, Y., et al. 1999. *Phys. Rev. Lett.* 82: 3472.
114 Landi, B. J., et al. 2002. *Nano. Lett.* 2: 1329.
115 Ahir, S. V., et al. 2005. *Nature Mater.* 4: 491.
116 Lu, S., et al. 2005. *Nanotechnology* 16: 2548.
117 Prentice, P. 1995. *Rheology and its role in plastics processing* (Rapra review report). Rapra Technology Ltd. 7: 3.
118 Liu, C., et al. 2003. *Polymer* 44: 7529.
119 Du, F., et al. 2004. *Macromolecules* 37: 9048.
120 Huang, Y. Y., et al. 2006. *Phys. Rev. B* 73: 125422.
121 Fan, Z., et al. 2007. *J. Rheol.* 51: 585.
122 Abbasi, S., et al. 2009. *Rheol. Acta.* 48: 943.
123 Bose, S., et al. 2009. *Compos. Sci. Technol.* 69: 365.
124 Siddiqui, N. A., Let al. (under review). *Compos. Sci. Technol.*
125 Mitchell, C. A., et al. 2002. *Macromolecules* 35: 8825.
126 Buldum, A., et al. 1999. *Phys. Rev. Lett.* 83: 5050.
127 Koratkar, N. A., et al. 2002. *Adv. Mater.* 14: 997.
128 Koratkar, N. A., et al. 2003. *Compos. Sci. Technol.* 63: 1525.
129 Suhr, J., et al. 2005. *Nature Mater.* 4: 134.
130 Salehi-Khojin, A., et al. 2008. *Compos. B* 39: 986.
131 Suhr, J., et al. 2008. *J. Mater. Sci.* 43: 4370.
132 Auad, M. L., et al. 2009. *Compos. Sci. Technol.* 69: 1088.
133 Koratkar, N. A., et al. 2005. *Appl. Phys. Lett.* 87: 063102.
134 Suhr, J., et al. 2005. Collection of technical papers—AIAA/ASME/ASCE/AHS/ASC Structures, Structural Dynamics and Materials Conference 3: 1550.
135 Zhou, X., et al. 2004. *Compos. Sci. Technol.* 64: 2425.
136 Rajoria, H., et al. 2005. *Compos. Sci. Technol.* 65: 2079.
137 Schadler, L. 2007. *Nature Mater.* 6: 257.
138 Kim, J. K., et al. 1998. *Engineered interfaces in fiber reinforced composites*, 1–100. New York: Elsevier.
139 Wagner, H. D., et al. 2004. *Mater. Today* 7: 38.
140 Desai, A. V., et al. 2005. *Thin-Walled Structures* 43: 1787.
141 Thostenson, E. T., et al. 2005. *Compos. Sci. Technol.* 65: 491.

142 Velasco-Santos, C., et al. 2005. *Compos. Sci. Technol.* 11: 567.
143 Ganesan, Y., et al. 2009. *JOM* 61: 32.
144 Cooper, C. A., et al. 2002. *Appl. Phys. Lett.* 81: 3873.
145 Liao, K., et al. 2001. *Appl. Phys. Lett.* 79: 4225.
146 Namilae, S., et al. 2007. *Comput. Model Engin. Sci.* 22: 189.
147 Liu, J. Q., et al. 2007. *Nanotechnology* 18: 165701.
148 Chen, X. L., et al. 2004. *Comput. Mater. Sci.* 29: 1
149 Wagner, H. D. 2002. *Chem. Phys. Lett.* 361: 57.
150 Schadler, L. S., et al. 1998. *Appl. Phys. Lett.* 73: 3842.
151 Paipetis, A., et al. 1999. *J. Compos. Mater.* 33: 377.
152 Cooper, C. A., et al. 2001. *Compos. A* 32: 401.
153 Hadjiev, V. G., et al. 2001. *Appl. Phys. Lett.* 78: 3193.
154 Zhao, Q., et al. 2004. *Phil. Trans. R Soc. Lond. A* 362: 2407.
155 Cui, S., et al. 2009. *Adv. Mater.* 21: 3591.
156 Ajayan, P. M., et al. 2000. *Adv. Mater.* 12: 750.
157 Wood, J. R., et al. 2001. *Compos. A* 32: 391.
158 Gao, S. L., et al. 2008. *Compos. Sci. Technol.* 68: 2892.
159 Qian, D., et al. 2002. *Appl. Mech. Rev.* 55: 495.
160 Qian, D., et al. 2003. *Compos. Sci. Technol.* 63: 1561.
161 Barber, A. H., et al. 2004. *Compos. Sci. Technol.* 64: 2283.
162 Barber, A. H., et al. 2005. *Adv. Mater.* 18: 83.
163 Deng, F., et al. 2007. *Compos. Sci. Technol.* 67: 2959.
164 Piggott, M. R. 1997. *Compos. Sci. Technol.* 57: 853.
165 Zhandarov, S., et al. 2001. *J. Adh. Sci. Technol.* 152: 205.

5

Application of CNT/Polymer Nanocomposites

5.1 Structural Application of CNT/Polymer Nanocomposites

Incorporation of carbon nanotubes (CNTs) into a polymer matrix produces a material with dramatically improved mechanical properties at low filler content, as discussed in Chapter 4. The use of CNT/polymer nanocomposites for structural components has several predictable impacts on current industries, especially in the automotive and aerospace fields. The main driving force behind the use of CNT/polymer nanocomposites in these industries is the ability to produce structural components with reduced weight-to-mechanical-performance ratios. The weight savings of the materials made of CNT/polymer nanocomposites arise from the excellent stiffness, strength, and other properties of CNTs that are significantly better than their metal counterparts. In addition, the use of nanocomposites in vehicle parts and systems is expected to shorten the manufacturing steps and time, as well as to enhance the environmental and thermal stabilities, and promote recycling.

5.1.1 Hybrid CNT and Fiber-Reinforced Polymer Composites

Fiber-reinforced polymers (FRP) are composite materials made of a polymer matrix reinforced with fibers. These fibers are usually made from glass, carbon, or aramid, while the polymer is usually an epoxy, vinylester, or polyester thermosetting plastics. FRPs are commonly used in the aerospace, automotive, marine, and construction industries. CNT/polymer nanocomposites can be employed as a modifier to enhance the performance of existing FRPs, either by *modifying the properties of fibers* or by *enhancing the properties of the polymer matrix*; in the latter case, CNT/polymer nanocomposites are employed as a matrix for FRPs.

Brittle fibers made from glass, graphite, alumina, and silicon carbide are widely used as reinforcements to produce high-performance FRPs with attractive mechanical and structural properties. One of the major issues concerning these fibers is that the measured strengths are significantly lower than their theoretical values. The strengths of these brittle fibers are influenced by the presence of surface and internal defects that are created during manufacturing and handling. Therefore, a thin film of sizings consisting of coupling agents, a film former, and other constituents is normally applied to glass fibers immediately after being drawn from molten glass[1]. These sizings aim to promote the fiber-matrix adhesion through the coupling effect, as well as to avoid damage during handling. It can also effectively reduce the stress concentration at the surface flaws by blunting the crack tips[2,3], providing a useful healing effect, thereby enhancing the strength of the fibers.

Various polymeric coatings as sizing or part of sizing have been extensively studied[1-4]. When an epoxy-based sizing was used, the distribution of the defects on the fiber surface became narrower than fibers without coating[5]. The idea of applying nano-filler-reinforced polymer coatings on glass fibers has recently been explored[6,7]. When a polymer sizing containing 0.2 wt% CNT was applied on the surface of alkali-resistant glass fiber, a significant improvement in fiber tensile strength was noted[7]; the CNTs in the polymer coating serve as the "bridges" at the defect tips on the fiber surface, which in turn delays the crack opening, as schematically described in Figure 5.1A. Siddiqui et al.[8] studied the effects of different fiber coatings and gauge lengths on tensile strengths of individual as well as bundled glass fibers. Three types of fibers, namely, fibers without coating, fibers with epoxy coating, and fibers with CNT/epoxy coating were employed. The results showed that the strengths of the fibers with epoxy coating were higher than those with coating for all of the gauge lengths studied and that the difference in strength became more pronounced with increasing gauge lengths (Figure 5.1B), which is a clear indication of reduction in notch sensitivity due to the fiber coating. It is thought that the epoxy coating filled the surface cracks, effectively increasing the crack tip radius and thus reducing the stress concentration at the defects. The results indicated that the smaller the effective crack length, the higher the fiber tensile strength. The

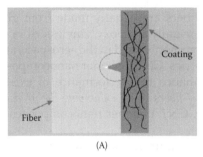
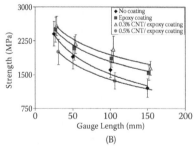

(A)   (B)

FIGURE 5.1

(A) Schematic illustration of a fiber with a CNT/epoxy nanocomposite coating, and (B) tensile strengths of single glass fibers as a function of gauge length for different coatings. Adapted from Siddiqui, N. A., et al. 2009. *Compos. A* 40: 1606.

tensile strengths of the fibers with a 0.3 wt% CNT/epoxy nanocomposite coating was even higher than those with the neat epoxy coating, indicating that significant synergy was provided by CNTs in the coating. This observation was common for all gauge lengths. It appears that the randomly dispersed CNTs within the nanocomposite coating offer a strengthening mechanism by bridging the surface microcracks. The crack-tip bridging effect promoted redistribution of the stresses around the surface cracks as far as the coating was intact with the glass fiber, thereby delaying the crack opening.

The examination of fracture surfaces of bundled fibers suggested that there was a transition in fracture mode between the samples with different coatings (Figure 5.2). The fiber bundles coated with neat epoxy failed predominantly by progressive fiber debonding and fiber pull-out, primarily in a *longitudinal splitting* mode, followed by breakage of fibers at multiple levels along the fiber direction (A and B in Figure 5.2). The extensively pulled-out fibers of different lengths on the fracture surface indicate that the damage was initiated from various locations along the interface. In sharp contrast, the fiber bundles coated with a CNT/epoxy nanocomposite failed transversely with relatively fewer debonding and fiber pull-outs, primarily in a

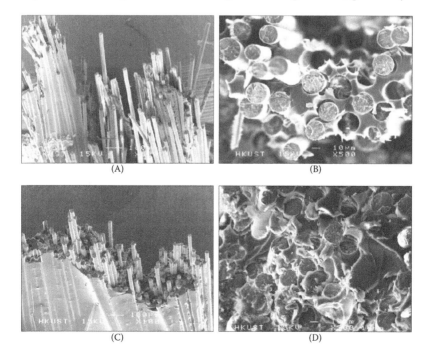

FIGURE 5.2
(A,B) SEM images of fractured surfaces of fiber bundles with neat epoxy coating and (C, D) with 0.3% CNT/epoxy nanocomposite coating. Adapted from Siddiqui, N. A., et al. 2009. *Compos. A* 40: 1606.

transverse fracture mode (C and D in Figure 5.2). In the latter samples, the interfaces between the fibers and the resin were undamaged even after failure, indicating that fibers were strongly bonded to the coating material. The failure was rather catastrophic compared to the former samples with the neat epoxy coating, suggesting that the presence of CNTs in the epoxy coating was mainly responsible for the transition in failure mode from longitudinal splitting to transverse fracture.

Another major application of CNT/polymer nanocomposites in FRPs is that the nanocomposites are employed as *matrix material* to enhance the performance of FRP composites. Hsiao et al.[9] developed a toughened nanocomposite consisting of multiwalled CNTs (MWCNTs) and epoxy as an adhesive to bond the graphite fiber/epoxy composite adhesive. Single lap joint samples were prepared and the average shear strengths were measured, showing a 46% increase by adding 5 wt% CNTs in the epoxy adhesive.

In FRPs, the formation and propagation of interlaminar cracks can lead to significant reductions in laminate strength and stiffness. The conditions that favor delamination can range from out-of-plane tensile loads to in-plane compressive loads, as well as local transverse low-velocity impact. Suppression of delamination is therefore of interest, particularly in primary structures made of FRPs. Grimmer et al.[10] reported that the addition of small volume fractions of MWCNTs, say 0.9 vol%, to the matrix of glass fiber-reinforced polymer (GFRP) can significantly reduce the rates of cyclic delamination crack propagation as well as both the critical and subcritical interlaminar fracture toughness values. In both the critical and subcritical cases, the degree of delamination suppression was most pronounced at low levels of applied cyclic strain energy release. High-resolution scanning electron microscopy (SEM) images of fracture surfaces suggested that the presence of CNTs at the delamination crack front slowed the crack growth through mechanisms like crack bridging, nanotube fracture, and nanotube pull-out. The relative proportion of CNT pull-out to CNT fracture was dependent on the applied cyclic strain energy in subcritical fracture. There was a shift in the fracture behavior of CNTs, which was responsible for the corresponding increase in the interlaminar fracture resistance compared to the composites without CNTs at low levels of cyclic strain energy release. Table 5.1 summarizes recent studies reporting improved mechanical properties of FRPs due to the incorporation of CNTs in the polymer matrix. The results confirmed that the fiber-dominated in-plane properties were not significantly affected by the CNTs, whereas the matrix-dominated properties, particularly the interlaminar shear strengths (ILSSs) were improved by 7–45%, as measured by the short beam shear (SBS) or the compression shear tests (CSTs)[11].

Indeed, FRPs modified by CNT/polymer nanocomposites have recently been commercialized. For example, Applied Nanotech Holdings, Inc. (ANI) has harnessed its expertise in CNT/polymer enhanced FRPs to develop a strengthened composite that can be used for wind turbine blades and other

TABLE 5.1 Improvement of ILSS of FRPs Due to CNT-Modified Matrices.

Fiber	Matrix	Type and Optimized Content of CNTs	Test Method	ILSS Improvement	Reference
Glass fiber	Epoxy	0.3 wt% amino-functionalized DWCNTs	SBS	20%	12
Glass fiber	Epoxy	1 wt% acid-functionalized MWCNTs	SBS	~7%	13
Glass fiber	Polyester and vinyl ester-epoxy resin	0.1 wt% amino-functionalzed SWCNTs	SBS	11%	14
Glass-fiber fabrics	Epoxy	0.3 wt% DWCNTs and MWCNTs	SBS	~5%	15
Woven glass fiber	Epoxy	1–2 wt% MWCNTs	CST	20.5% (1 wt%) 33% (2 wt%)	16
Woven glass fiber	Vinyl ester	0.1 wt% pristine and functionalized SWCNTs	SBS	20–45%	17
Woven carbon fiber	Epoxy	CNTs are selectively deposited on fiber	SBS	30%	18

applications with long lifetime requirements[19]. Fiberglass turbine blades used for wind energy need to withstand a range of environmental conditions over many years. Current wind blade research efforts focus on how to achieve longer lifetimes with lower-weight materials and blade designs to reduce overall cost and increase efficiency. ANI is exploring ways to lower the blade weight with CNT-strengthened materials and thus to achieve the desired objectives of the industry. The early stage research and development (R&D) in this area has shown some dramatic results[19]. With the development of this technique, more and more commercialized products will emerge in the market for many different applications.

5.1.2 Automobile Applications

The automotive industry is a material-intensive industry. A wide variety of metals, fillers, and plastics are used today to meet the requirements of specific service conditions. The ultimate drivers in materials selection are cost and performance. Materials are selected by identifying the best cost/performance ratio needed to meet the requirements of the application. This reality has driven the development of numerous approaches to enhance the properties of conventional materials[20]. Examples of these approaches include structural plastics, alternative metals and alloys, reinforcing fillers, and glass/carbon FRPs. Each of these approaches has limitations[20–23]. For example, structural plastics often require postforming modifications of the surface and long cyclic life and are more expensive. Lightweight metals and their alloys are finding increased use, primarily in noncosmetic structures, for weight reduction. These metals, however, have the same processing limitations as steel and iron, and they usually add cost. Reinforcing fillers such as talc, mica, and calcium carbonate, for example, introduce higher stiffness while also increasing the weight and melt viscosity, and decreasing the toughness, optical clarity, and surface quality. Glass-fiber reinforcements provide higher stiffness with a corresponding increased difficulty of fabrication and cost. These traditional reinforcements and fillers must be used at high loading levels to increase the modulus and improve the dimensional stability, thus compromising weight, toughness, and surface quality[20–23].

In contrast to traditional microscale fillers, nanofillers such as CNTs and nanoclays[22] are expected to be effective in improving the desired properties with a low content, usually less than 5% by weight, producing only a minor increase in material cost. Even so, they provide significant improvements in strength, modulus, thermal stability, dimensional stability, surface hardness, heat-distortion temperature, mar resistance, and barrier properties. Nanoscale reinforcement would make the design of parts and systems more cost effective, allowing the components traditionally made from metals and glass to be replaced by polymer nanocomposites for production of fuel-efficient, higher-quality, and more durable vehicles.

There have been significant R&D activities involving CNT/polymer nanocomposites for automotive applications: major focuses were to enhance the prime structural properties and the robustness of new composite materials as well as their automation and large-scale production. Figure 5.3 shows the potential structural applications of CNT/polymer nanocomposites in the automobile industry. Several big names in the automobile industry, such as Toyota and GM, have been exploring the use of CNT/polymer nanocomposites to replace current structural composites for automobiles in collaboration with Unidym, Inc. and Batelle[24]. High-performance carbon-fiber composites containing CNTs were developed, where CNT/thermosets were injected and molded by resin transfer molding (RTM) processes through the compression molding of nanostructured sheet molding compound formulations.

The realized structural components made from CNT/polymer nanocomposites are automobile bumpers[25]. The bumper consisting of 1–5 wt% CNT and polycarbonate exhibited good mechanical properties and lower weight than the standard fiberglass bumpers in which 30 wt% or more of fiberglass was needed. The bumpers showed their multifunctionalities, possessing good mechanical properties, low weight and electrical conductivity. The conductivity of the bumper allowed the direct electrostatic spray application

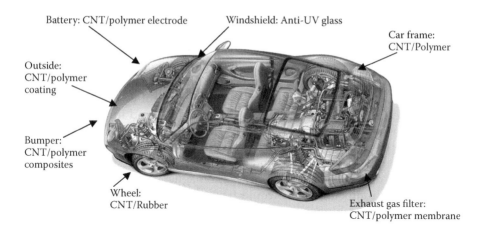

Battery: CNT/polymer electrode

Windshield: Anti-UV glass

Car frame: CNT/Polymer

Outside: CNT/polymer coating

Bumper: CNT/polymer composites

Wheel: CNT/Rubber

Exhaust gas filter: CNT/polymer membrane

FIGURE 5.3
Possible structural applications of CNT/polymer nanocomposites in an automobile.

of base and clear coats, eliminating the need for an additional primer coat prior to painting. Additional advantages were savings in paint consumption and reductions in volatile emissions from paint lines. In another application, the Degussa-Hüls chemical company[26] has developed a CNT/nylon nanocomposite for tubing in the flexible portion of the fuel lines in 70% of the cars made in North America.

5.1.3 Aerospace Applications

Polymer-based composites have been used as critical components in aircraft. For example, the new Boeing 787 Dreamliner features lighter-weight construction due to the employment of about 50% carbon and glass fiber–reinforced composites with 80% of the craft being made from composites by volume. Composites are used on the fuselage, wings, tail, doors, and interior, and so on[27]. To demonstrate potential aircraft application of CNT/polymer nanocomposites, O'Donnell et al.[28] conducted a mass analysis study on a CNT-reinforced polymer Boeing 747-400 aircraft without including any modifications to the geometry or design of the airframe. The primary structural aluminum material was replaced entirely by CNT/high-density polyethylene (HDPE) nanocomposites. The results achieved an average 17.32% weight reduction in the low initial takeoff mass category. The average fuel savings for all CNT-reinforced airframes was about 10%. In summary, this type of analysis provides insight into the ultimate advantages of CNT/polymer nanocomposites for aerospace applications.

Organic polymers with uniformly dispersed CNTs may enable polymer materials to withstand the harsh space environment and may be used for the purpose of critical weight-reduction on current and future space systems. Lincoln et al.[29] speculated that launch vehicles would greatly benefit from appropriately designed nanocomposites that could provide improved barrier properties and gradient morphologies for composite cryogenic fuel tanks. Self-frigidizing and self-passivating nanocomposite materials could be used to construct space vehicle components that are both highly resistant to space-borne particles and resistant to degradation from electromagnetic radiation, while reducing the overall weight of the spacecraft[30]. All of these characteristics and advantages can be found in CNT/polymer nanocomposites.

CNT/polymer nanocomposites also offer a unique opportunity for improved durability of physical and structural properties, such as the coefficient of thermal expansion and antielectromagnetic radiation, in an interplanetary environment, which would be especially useful in constructing large apparatuses in human space exploration beyond the low-earth orbit[30,31]. Lake[32] also indicated that CNT/polymer nanocomposites had a high potential for low-cost and large-volume production for aerospace applications. However, the number of papers dedicated to this subject is very limited. This does not necessarily mean lack of interest, but rather lack of immediate, direct commercial and military implications, and consequently secretive pursuit of this kind of research. In summary, it is believed that the fabrication and realization of CNT/polymer nanocomposites in the aerospace industry are under rapid development, and only a limited number of applications have so far been achieved.

5.1.4 Other Structural Applications

There are still many opportunities to employ CNT/polymer nanocomposites for different structural applications such as materials used in civil engineering and public security. For example, polymer materials with high ductility are usually soft, and the stiffer the materials, the less ductile they are. Introduction of CNTs into some thermoplastics can enhance stiffness without sacrificing ductility. Recently, researchers at the Hong Kong University of Science and Technology (HKUST) have developed a new technology that can greatly enhance the ballistics-proof strength of fibers made from ultra-high molecular-weight polyethylene (UHMWPE) by adding CNTs into the polymer[33]. This new technology is expected to pave the way for new structural applications of UHMWPE, such as more comfortable and effective bulletproof vests and extra-durable nautical ropes.

The multidisciplinary nature of CNT/polymer nanocomposites inevitably makes the boundary between the structural and functional applications increasingly more indistinctive. Indeed, many of the functional requirements of CNT/polymer nanocomposites are derived while designing and fabricating structural components, and this aspect will be discussed in the following section.

5.2 Functional Application of CNT/Polymer Nanocomposites

5.2.1 Conducting Films and Coatings

Conducting films and coatings are essential materials for many device applications, such as displays, touch panels, plastic solar cells, and so on. They should act both as windows for light to pass through to the active material beneath where carrier generation occurs and as Ohmic contacts for carrier transport out of the photovoltaic device. Transparent materials possess bandgaps with energies corresponding to wavelengths that are shorter than the visible range. As such, photons with energies below the bandgaps are not collected by these materials and thus visible light passes through.

Conducting films and coatings have been fabricated from both inorganic and organic materials. Inorganic films are typically made up of a layer of conducting oxide, generally in the form of indium tin oxide (ITO), fluorine-doped tin oxide (FTO), and zinc oxide. However, the performance of inorganic/polymer films is generally low compared to that on glass; for example, conductivities of ITO on polymers are much lower than those on glass. In addition, cracks appear after repeated bending or strain, and this material is not resistant to acid[34]. Due to these technical limitations, polymer-based conducting films were recently developed to meet wider applications. CNTs exhibit excellent mechanical and electrical properties, while polymer matrices provide good flexibility, high transparency, easy processing, and low cost. Thus, the combination of these two materials presents a new direction to develop new conducting films and coatings to replace those in current use.

There are basically three main approaches to fabricating CNT/polymer conducting films—solution blending, melt blending, and in-situ polymerization[35]. Two basic properties of CNT/polymer films have to be carefully considered in fabrication: one is the electrical conductivity, and the other is the mechanical properties. In some specific applications, the optical and thermal properties of films are also important. As discussed in Chapter 4, several factors, such as CNT aspect ratio, dispersion state, interfacial interaction with the polymer matrix, govern the properties of CNT/polymer nanocomposites, and these factors are equally important for the properties of thin films. Peng[36] reported a novel technique to produce CNT/polymer nanocomposite films with highly aligned nanotubes for applications as flexible conductors for optoelectronic devices. The resulting nanocomposite films showed high optical transparency, robust flexibility, and much improved electrical conductivities compared to the films obtained by other approaches. In another study[37], CNT/polyethylene thin films were fabricated by swelling the polyethylene in a CNT/tetrahydrofuran solution, followed by the infiltration of nanotubes. These thin films, typically 250 nm thick, displayed conductivities of up to 66 S/m, depending on the CNT content. This value is equivalent to a sheet resistance as low as 50 kΩ/Sq, with an optical transparency of 80%. The reasonably low sheet resistance and high transparency of these films make them ideally suited to various optoelectronics and display applications. There is a large volume of literature regarding the processing, optimization, and performance evaluation of thin films and coatings based on CNT/polymer nanocomposites, and Table 5.2 summarizes the most recent studies in this field. However, the commercialization of these conducting thin films and coatings is still in its infant stage, and further research and development efforts are warranted to satisfy the market demands to replace the existing materials before widespread applications are realized.

5.2.2 Electromagnetic Interference Shielding

Electromagnetic interference (EMI) shielding refers to the reflection and/or adsorption of electromagnetic radiation by a material, which thereby acts as a shield against the penetration of the radiation through it. Lightweight EMI shielding is needed to protect the workspace and environment from radiation coming from computers and telecommunication equipment as well as for protection of sensitive circuits. Compared to conventional metal-based EMI shielding materials, electrically conducting polymer nanocomposites have gained popularity recently because of their light weight, resistance to corrosion, flexibility, and processing advantages[46].

The EMI shielding efficiency of a polymer nanocomposite depends on many factors including the intrinsic conductivity of the filler, its dielectric constant, and aspect ratio. The small diameter, high aspect ratio, high conductivity, and mechanical strength of CNTs make them an ideal filler material for producing conductive composites for high-performance EMI shielding applications. The studies of single-walled CNT (SWCNT)/polymer nanocomposites[46] indicate that SWCNTs can be used as effective lightweight EMI shielding materials that require a high DC conductivity. Nanocomposites with greater than 20 dB shielding efficiency were obtained, and the highest EMI shielding efficiency (SE) was obtained from 15 wt% SWCNTs, reaching 49 dB at 10 MHz and 15–20 dB in the frequency range of 500 to 1.5 GHz.

Yang et al.[47] developed CNT/polystyrene (PS) nanocomposite foam for EMI shielding applications. The EMI shielding effectiveness provided by this composite was around 20 dB at CNT loading of 7 wt%, which is comparable to that of commercialized composites used as shielding media against electromagnetic radiation. The CNT/PS composites were found to be more effective in providing EMI shielding than carbon nanofiber/PS nanocomposites at the same filler content. The EMI shielding mechanisms of CNT/polymer nanocomposites were analyzed experimentally in a recent study[48]. MWCNT/polypropylene (PP) nanocomposites with varying thicknesses of 0.34, 1.0, and 2.8 mm and CNT content were fabricated. Three mechanisms were identified for EMI shielding: reflection, absorption, and multiple reflection (Figure 5.4A). The shielding by absorption of MWCNT/PP nanocomposites

TABLE 5.2 Fabrication and Application of CNT/Polymer Thin Films and Coatings.

CNT Type	Matrix	Fabrication Process	Function	Advantages	Reference
MWCNT	Polyethylene	Layer-by-layer technique	Conducting film	High conductivity and transparency	37
MWCNT	Polypyrrole	Gas-phase polymerization	Conducting film	High conductivity and uniformity, excellent thermal stability	38
MWCNT	Poly(3-exylthiophene)	Spin casting	Conducting film	Good processability and intense PL in the orange	39
MWCNT	Polybenzazole	Solution casting and dipcoating	Conducting film	Improved electrical, mechanical properties, excellent thermal stability	40
MWCNT	Polyetherimide	Solution casting	Multifunctional coating	Improved electrical conductivity and excellent thermal stability, good processability	41
SWCNT	Poly (vinyl alcohol) Poly (sodium 4-styrene–sulfate)	Layer-by-layer assembly	Conducting coating	High conductivity and transparency, excellent mechanical properties	42
SWCNT	Polycarbonate or polystyrene	Drop casting or spin coating	Conducting film	Low CNT content, excellent processability	43
SWCNT	Poly(3,4-thylenedioxy-thiophene)	Microcontact printing	Conducting film	Low CNT content, High conductivity and transparency, good processability	44
SWCNT	UHMWPE	Dipcoating	Protective coating	Excellent friction and wear resistance as boundary lubricants	45

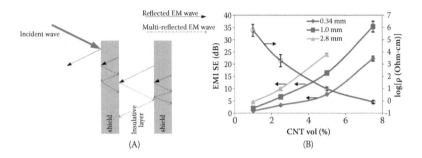

FIGURE 5.4
(A) A schematic showing the mechanism of EMI, and (B) electrical resistivity and EMI SE of MWCNT/PP nanocomposites as a function of CNT content and shielding plate thickness. Adapted from Al-Saleh, M. H., et al. 2009. *Carbon* 47: 1738.

depends on the distance between the MWCNT particles and/or the electrical resistivity of the nanocomposite. Shielding by reflection was found to depend on the conductivity and thickness of the nanocomposite and the CNT content (Figure 5.4B). A multisurface shield might significantly enhance the overall EMI SE if multiple reflection can be minimized. For nanocomposites with a thickness of 0.34 mm, absorption is the primary shielding mechanism, followed by shielding by reflection. The theoretical analysis showed that multiple reflection within the MWCNT internal surfaces might have a negative influence on the overall EMI SE because the MWCNT diameter was orders of magnitude smaller than the skin depth. Multiple reflection between external surfaces of MWCNTs also decreased the overall SE, but its influence was smaller than that between the internal surfaces.

5.2.3 Sensors and Actuators

CNTs are excellent materials for use as electrodes in sensors because their electronic transport and thermopower, or the voltages between junctions caused by interjunction temperature differences, are very sensitive to substances that affect the amount of injected charge[49]. The main advantages of CNT sensors are the minute size of the nanotube sensing element and the correspondingly small amount of material required for a response. Li et al.[50] provides a most comprehensive review of the subjects covering sensors and actuators made from CNTs and CNT nanocomposites for different applications. A wide range of sensor architectures, including CNT paste electrodes, glass electrodes modified by CNTs, and metal nanoparticle–modified CNT electrodes, have been developed to monitor the existence of gases, biological substances, drugs, and so on[51]. These sensors have an excellent sensitivity at room temperature due to a high sensitivity to changes in the electrical conductivity upon the adsorption of sensing materials. Despite such advantages, however, their applications are limited by a long recovery time and a complex fabrication process. In addition, to maximize the sensing performance, the CNT-based sensors must be semiconducting. The presence of both metallic and semiconducting CNTs in conventional CNT samples reduces the reproducibility and yield of these devices[51]. As discussed in Chapter 4, the combination of CNTs with polymers not only provides higher stability and better uniformity, but also endows the nanocomposite with multifunctional properties, such as enhanced electrical and thermal properties and good processability. Table 5.3 summarizes recent developments of sensors based on CNT/polymer nanocomposites, and highlights the principles and suitable sensing materials for different types of sensors.

Recent reports have proposed the use of CNT/polymer nanocomposites as vapor sensors[52,53]. The principle of these sensors is that the polymer matrix can be swollen due to the adsorption of solvent when exposed to a sensing medium, which in turn results in a change in electrical conductivity of nanocomposites. CNT/poly(methyl methacrylate) nanocomposites were employed to detect biohazard materials and vapors. The results showed that the resistance change of nanocomposites was governed by the concentration of targeting vapors diffused into the nanocomposite, as a reflection of the transition of conducting networks from direct contact to tunneling in CNT junctions. The use of sensors in a wireless sensing system showed a large differential phase shift, indicating potential application for remote monitoring of biohazard vapors in real time[52]. Wei et al.[53] also demonstrated that the aligned CNT/polymer nanocomposite displayed similar changes in resistance when subjected to thermal or some chemical vapors, such as ethanol,

TABLE 5.3 Application of CNT/Polymer Nanocomposites for Various Sensors.

Sensor Type	Principle	Sensing Materials
Biosensor	Fast electron transfer and ion mass transfer between nanocomposite and target materials	Biomolecules, such as glucose, DNA, proteins, drugs, enzymes, etc; toxic heavy metal ions in liquids, such as Pb, Cr, Pd, Cu, etc; pH value of acid/base solutions
Flow sensor	Generation of voltage along the direction and velocity of the flowing	Liquids with low viscosities
Strain sensor	Electrical conductivity change due to the mechanical load of nanocomposites	Bulky composites, FRPs, etc.
Vapor sensor	Electrical conductivity change due to the volume change of nanocomposites	Gases: N_2, H_2, O_2, NH_3, CO, CO_2, etc.; volatile liquids: H_2O, H_2O_2, toxic organic solvents, etc.
IR sensor	Photoresponse change of insulating polymer matrix	Living creatures that radiate heat energy in environment
Thermal sensor	Thermal expansion of nanocomposites	Temperature change in liquid or gas

tetrahydrofuran, toluene, hexane, and so on. Again, the resistance changes were attributed to expansion of the polymer matrix, indicating potential application of CNT nanocomposites as mechanical sensors.

CNT/polymer nanocomposites can also be employed as a mechanical strain sensor, an idea that has attracted significant research interest. Zhang et al.[54] reported a prototype CNT/polycarbonate nanocomposite utilized as a strain sensor, and showed that when subjected to linear and sinusoidal dynamic strain inputs, the instantaneous change in the electrical resistance of the nanocomposite responded in a manner similar to a strain gauge. The sensitivity of the nanocomposite sensor was measured to be 3.5 times higher than a traditional strain gauge, suggesting its potential applications for self-diagnostics and real-time health monitoring of structures. In another study, Kang et al.[55] used MWCNT/poly(methyl methacrylate) nanocomposite films for strain sensing and examined their static and dynamic behaviors. The principle of this sensor is that CNTs can be considered long continuous neuron sensors in nanocomposites, and these neurons can be connected to form conducting networks, like the neural system in a human body. The neural system can exist in the form of a grid attached to the surface of a structure to make a sensor network, thus enabling structural health monitoring. The results showed that the sensitivity of this sensor matched well with the vibration measured using a laser displacement sensor.

One of the common features among many reports on vapor or strain sensing by CNT/polymer nanocomposites is that a change in volume due to chemical, thermal, or mechanical loading results directly in the changes in electrical resistance of nanocomposites. Indeed, this concept is a continuation of employing electrical techniques as a noninvasive way to monitor damages in carbon fiber–reinforced polymers (CFRPs) under static and dynamic loads[56–59]. Because carbon fibers are conductive, fiber fracture results in the change in electrical resistance of composites. However, this approach does not give much insight into matrix-dominated fracture mechanisms that affect durability and fatigue life, and is not applicable to composites with nonconducting fibers, such as glass or aramid fibers. Fiedler et al.[60] proposed the concept of using CNT/polymer nanocomposites for both strain and damage sensing of structural materials, and Thostenson and Chou further developed this technique to fabricate hybrid composites reinforced with conventional micron-sized fibers and CNTs[61], making composites with an in-situ sensing capability. By combining these reinforcement fillers of different scales, CNTs can penetrate the matrix-rich areas between fibers in individual bundles as well as between adjacent plies (Figure 5.5A) and can achieve a percolating nervelike network of sensors throughout the arrays of fibers in a composite. These percolating networks of CNTs were sensitive to initial stages of matrix-dominated failure, being able to detect damage in situ (Figure 5.5B). Through experiments designed to promote different failure modes, the nature and progression of damage were identified.

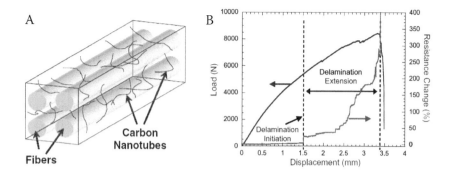

FIGURE 5.5
(A) Three-dimensional model showing the penetration of nanotubes throughout a fiber array, and (B) the corresponding load/displacement and resistance responses of a composite with the center ply cut to initiate delamination. Adapted from Thostenson, E. T., et al. 2006. Adv. Mater. 18: 2837.

The sensor applications of CNT/polymer nanocomposites are not limited to the aforementioned types. Indeed, any changes arising from surrounding environmental conditions as well as thermal, electrical, biological, mechanical, and optical variations will result in the corresponding changes in the physical and mechanical responses of nanocomposites, all of which can be employed for sensor applications.

In addition to utilizing the electrical and mechanical behaviors of CNTs in sensing applications, there is a growing interest in utilizing CNT/polymer nanocomposites as actuators[50]. An actuator is a mechanical device that takes energy, usually created by air, electricity, light, or liquid, and converts it into some kind of motion. Photomechanical actuators convert light into a mechanical motion, and the principles and applications of these actuators based on CNT/polymer nanocomposites are discussed in Section 4.3.5. In addition to this kind of actuator, the electromechanical behavior of CNT/polymer nanocomposites can also be employed to fabricate electromechanical actuators for nano-electro-mechanical systems (NEMSs) and nanoscale switches[50]. Electromechanical actuators based on SWCNT sheets were shown to generate higher stresses than natural muscles and higher strains than high-modulus ferroelectrics[62]. These actuators are assemblies of billions of individual nanoscale actuators, and the actuation mechanism does not require ion intercalation and can be operated at a low voltage to generate large actuation strains. Levitsky et al.[63] created layered SWCNT/Nafion nanocomposites by applying a thin coating of SWCNTs to the surfaces of the membrane as electrodes and investigated their electrochemical actuation behavior. The results showed that the actuators gave a large mechanical response at low voltages without the use of an electrolyte. They concluded that the large surface area of CNTs played a role in enhancing actuation, and CNTs themselves were also responsible for some of the actuation response.

Further research on similar nanocomposites showed that the dispersed CNTs served as an embedded network of electrodes throughout the polymer matrix and enhanced the electro-osmotic effect, thus enhancing the actuation response. The best performance in terms of maximum stress and stress rate occurred with uniform distributions of CNTs[64,65].

Yun et al.[66] developed ply actuators where an epoxy layer was cast onto the surface of a random CNT paper. The actuator was tested in a liquid electrolyte to characterize the actuation behavior. Figure 5.6 (A and B) shows the behavior of the actuator after applying 10 V for a period of minutes. It is clear that the displacement became very large, the actuator curled, and this was not reversible. However, this kind of actuator has some inherent drawbacks, such as reagglomeration, random distribution, and low aspect ratio of CNTs in the epoxy matrix. It was proposed that incorporation of vertically aligned CNT arrays into the CNT plies to build a 3-D CNT/polymer nanocomposite (Figure 5.6C) could further improve the performance of the actuator. This has been confirmed by a recent study in which a nanocomposite actuator consisting of vertically aligned CNTs and Nafion was developed[67], showing a significantly improved ion transport speed in the CNT alignment direction. The high elastic anisotropy, arising from the high modulus and volume fraction of aligned CNTs, enhanced the actuation strain while reducing the undesirable direction strain. More than 8% actuation strain at 4 volts with less than one second response time has been achieved. This smart structural material can have potential applications that range from robotic surgical tools to structures that are able to change shapes.

Since the first paper published in this field in 2000[68], significant progress has been made by employing CNT/polymer nanocomposites for actuator applications. Table 5.4 summarizes recent studies in this field, highlighting the structure and the corresponding performances of these actuators by employing different CNTs and polymer matrices.

FIGURE 5.6
Response of a CNT/epoxy actuator (A) before and (B) after applying a voltage; and (C) a schematic showing the concept of multilayer nanocomposite actuator. Adapted from Yun, Y. H., et al. 2005. *Smart Mater. Struct.* 14: 1526.

TABLE 5.4 Application of CNT/Polymer Nanocomposites for Various Actuators.

CNT Type	Matrix	Structure	Performance	Reference
Vertically aligned CNTs	Nafion	—	Enhanced actuation strain with reduced undesirable direction strain. > 8% actuation strain under 4 volts, quick response time	67
MWCNT	Poly(vinylidene fluoride-trifluoroethylene) copolymer	Thin film	Giant lateral electrostriction, high and fast strains of up to 4% within an electric field was achieved by electrostriction	68
DNA-wrapped MWCNTs	Polypyrrole	Composite fiber	Excellent actuation stability with an expansion and contraction of –4.4% under a low voltage (±1 V). Fast redox response under different chemical conditions	69
MWCNT bucky paper	Polyvinylidene difluoride	Sandwich	Wide actuation stability without faradaic reactions	70
MWCNTs	Poly(vinyl alcohol)/poly(2-acrylamido-2-methyl-1-propanesulfonic acid) blend	Cantilever	Low CNT content, generation of large blocking tip force	71
MWCNTs	Polydimethylsiloxane	Nanocomposite deposited on glass substrate	Large actuation behavior under low voltage, stable under different applied voltage	72
MWCNTs	Polyurethane	Bulky composites	Excellent durability under different temperatures	73

5.2.4 Applications for Energy Storage

CNT/polymer nanocomposites have many useful properties that can be employed as materials for energy storage and conversion to energy dissipation (Figure 5.7). Among many energy storage applications, CNT/polymer nanocomposites have been utilized to develop high-performance supercapacitors. Supercapacitors, also known as electrochemical capacitors, are important devices in energy storage and conversion systems and are considered for a variety of applications, such as electric vehicles, support for fuel cells, uninterruptible power supplies, memory protection of computer electronics, and cellular devices. Traditionally, three types of materials can be used for capacitor applications: carbon, transition metal oxides, and conducting polymers. Polymer-based supercapacitors represent an interesting class, thanks to the combination of high capacitive energy density and low material cost. However, this kind of supercapacitor has some disadvantages, such as shorter cycle life and slow ion transport, because the redox sites in the polymer backbone are not sufficiently stable for many repeated redox processes[74,75]. The introduction of CNTs into a polymer matrix improves the electric conductivity as well as the mechanical properties of the polymer matrix[76–78], while possibly providing an additional active material for capacitive energy storage.

Using MWCNTs coated with polypyrrole (PPY) as the active electrode for a supercapacitor assembly, Frackowiak et al. reported an increase in specific capacitance from 50 to 180 F/g, demonstrating a synergy between the two components of the nanocomposites[79,80]. The open mesoporous network of these conducting nanotubular materials allows an easy access of ions to the electrode/electrolyte interface and a more effective contribution of the pseudo-faradaic properties of PPY, as well as long durability (over 2000 cycles). The performance of supercapacitors can be further improved either by employing different

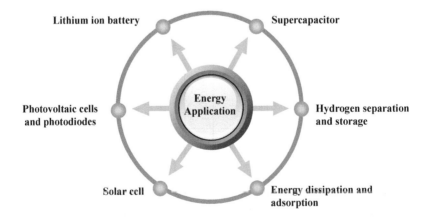

FIGURE 5.7
Typical applications of CNT/polymer nanocomposites for sustainable energy generation and energy storage.

polymer matrices, such as polyethylenedioxythiophene, polyaniline, and polythiophene derivatives, or by using CNTs with different characteristics through purification or having different numbers of multiwall layers, and so on.

Another energy storage application of CNT/polymer nanocomposites is to develop high-performance lithium ion batteries (LIBs). A typical LIB consists of a negative electrode (i.e., an anode typically made of graphite), a positive electrode (i.e., a cathode made typically of $LiCoO_2$), and a lithium–ion conducting electrolyte. When the cell is charged, Li ions are extracted from the cathode, move through the electrolyte, and are inserted into the anode. Upon discharge, the Li ions are released by the anode and taken up again by the cathode, and the electrons pass around the external circuits. LIBs possess many advantages over traditional rechargeable batteries, such as lead acid and Ni-Cd batteries; they include high voltage, high energy-to-weight ratio, long cyclic life, no memory effect, and slow loss of charge when not in use[81–83]. For these reasons, LIBs are currently the most popular type of battery for portable electronic devices and are growing in popularity for defense, automotive, and aerospace applications. Although LIBs have achieved remarkable commercial success, a major disadvantage of LIBs is their low power density, which is the consequence of high polarization, especially at high charge–discharge rates. Slow diffusion rates of Li ions across the electrolyte–electrode interface, and poor electrical and thermal conduction in the electrodes are mainly responsible for the high polarization. To overcome these shortcomings, it is important to find new electrode materials that can provide a higher surface area and a shorter diffusion path for ionic transport and higher electrical and thermal conduction.

CNT/polymer nanocomposites offer an exciting alternative to the standard materials traditionally used for the fabrication of LIBs. Baibarac et al.[84] fabricated CNT/poly(N-vinyl carbazole) (PVC) nanocomposites using electrochemical polymerization of N-vinyl carbazole on CNT films. The nanocomposites performed better than the PVC material alone, as a positive electrode of rechargeable LIBs, with higher specific discharge capacities of 45 and 115 mAh/g after 20 cycles by employing SWCNT/PVC and MWCNT/PVC nanocomposites, respectively. Aligned CNTs were also employed as an innovative structure to further improve the electrochemical performance of CNT/polymer nanocomposites used in LIBs. Chen et al.[85] developed an approach for preparing a nanocomposite consisting of aligned CNTs and Poly(3, 4-ethylenedioxythiophene) (PEDOT) where the CNTs were held together by a conducting PEDOT layer. The nanotubes remained vertically oriented and protruded from the conducting polymer layer with 90% of the tube length being exposed. This novel "freestanding" nanocomposite electrode was characteristically lightweight, flexible, highly conductive, and mechanically robust, and could be easily fabricated into a rechargeable battery without using a metal substrate or binder. The weight of the electrode was reduced significantly from the conventional electrode, which is made by coating the active material onto the metal substrate. In addition, the capacity of the nanocomposite electrode was

TABLE 5.5 Other Functional Applications of CNT/Polymer Nanocomposites.

Application	Exploited Properties of Nanocomposites (due to the advantages arising from CNTs)	Principle or Advantages
Thermal interface materials	High thermal conductivity, low thermal shrinkage	CNT–filled nanocomposites fill the gaps between thermal transfer surfaces (such as between microprocessors and heatsinks), thus increase the efficiency of heat transfer.
High-resolution printable conductor	High electrical conductivity and good processability of nanocomposites	Addition of CNTs in polymers, e.g., polyaniline doped with dinonylnaphtalene sulfonic acid, creates a highly conducting three-dimensional percolating network. The thin nanocomposite films are printable at a high resolution while maintaining appropriate conductivity.
Schottky diodes	Enhanced thermal and mechanical stability as well as overall conductivity of devices	Well-dispersed CNTs in a polymer matrix act as nanometric heat sinks, preventing the build-up of large thermal effects and thus reducing material and device degradation.
Field-effect transistors	Low-cost materials and easy processing procedures to fabricate devices with good device performance	CNT/polymer nanocomposites possess strong gate-channel coupling with improved device properties. CNTs can be easily doped using different concentrations of electron acceptor mixed with polymers. The transport type of the devices can be easily controlled through doping.
Optical limiter	Excellent optical limiting properties of nanocomposites	CNTs in nanocomposites block the incident light effectively at high incident energy densities or intensities through nonlinear scattering, absorption, refraction, and electronic absorption.
Tissue engineering	Enhanced strength and toughness, lightweight, not biodegradable, good compatibility with tissues	CNT/polymer nanocomposites behave like an inert matrix on which cells can proliferate and deposit new living material; Good in-vivo biocompatibility and positive tissue response.
Biomedical applications	Enhanced mechanical properties, good structure design and processability, etc.	Excellent biocompatibility and multifunctionalities in drug, dental, and ophthalmic applications.

50% higher than that observed for a freestanding SWCNT paper. This study has important implications for the use of aligned CNT/conductive polymer composites as a new class of electrode material in developing flexible and high-performance rechargeable LIBs.

CNT/polymer nanocomposites have also been widely employed to fabricate photovoltaic devices for energy conversion[51,86]. The primary step required for producing photovoltaic devices is an ultra-fast photo-induced electron transfer reaction where the polymer matrix acts as the electron donor and CNTs as the electron acceptor at the donor–acceptor interface, resulting in a metastable charge-separated state. In the case of the oligo(phenylenevinylene)/C_{60} nanocomposite, the quantum efficiency of this step was assumed to be close to one. However, the overall conversion efficiency of these devices was limited by the carrier collection efficiency, which was greatly influenced by the morphology of the active film. Optical and photovoltaic properties of photovoltaic devices made with CNT/polymer nanocomposites confirmed significantly improved quantum efficiency due to the introduction of CNTs, along with enhanced thermal stability. There have been more than 1000 papers published in this area since 2000, and Baibarac et al.[51] reviewed the recent progress on the development of CNT/polymer nanocomposites for applications in photovoltaic cells, photodiodes, solar cells, and so on.

5.2.5 Other Functional Applications

In addition to the aforementioned applications, CNT/polymer nanocomposites can be used as multifunctional components to develop new materials used in many different fields, especially in biomedical and electronic devices. Table 5.5 summarizes these applications, highlighting the properties and principles behind the use of these materials for the specific areas.

While much work still needs to be done, polymer/CNT nanocomposites offer immense potential for applications in many fields, and there is no doubt that this will have significant influence on the development of composite science and technology for many years to come.

References

1 Kim, J. K., et al. 1998. *Engineered interfaces in fiber reinforced composites*, 179. New York: Elsevier.
2 DiBenedetto, A. T., et al. 1989. *Polym. Eng. Sci.* 29: 543.
3 Zinck, P., et al. 1999. *J. Mater. Sci.* 34: 2121.
4 Loewenstein, K. L. 1983. *The manufacturing technology of continuous glass fibers*, 1–52. New York: Elsevier.
5 Ahlstrom, C., et al. 1995. *Polym. Compos.* 16: 305.

6 Gao, S. L., et al. 2007. *Acta Mater.* 55: 1043.
7 Gao, S. L., et al. 2008. *Compos. Sci. Technol.* 68: 2892.
8 Siddiqui, N. A., et al. 2009. *Compos. A* 40: 1606.
9 Hsiao, K. T., et al. 2003. *Nanotechnology* 14: 791.
10 Grimmer, C. S., et al. 2010. *Compos. Sci. Technol.* 70: 901.
11 Qian, H., et al. 2010. *J. Mater. Chem.* 20: 4751.
12 Gojny, F. H., et al. 2005. *Compos. A* 36: 1525.
13 Qiu, J. J., et al. 2007. *Nanotechnology* 18: 275708.
14 Seyhan, A. T., et al. 2008. *Eng. Frac. Mech.* 75: 5151.
15 Böger, L., et al. 2008. *Compos. Sci. Technol.* 68: 1886.
16 Fan, Z. H., et al. 2008. *Compos. A* 39: 540.
17 Zhu, J., et al. 2007. *Compos. Sci. Technol.* 67: 1509.
18 Bekyarova, E., et al. 2007. *Langmuir* 23: 3970.
19 Applied Nanotech Holdings, Inc., http://www.appliednanotech.net/tech/cnt_fiberglass.php (accessed January 2010).
20 Takemori, M. T. 1979. *Polym. Eng. Sci.* 19: 1104.
21 Giannelis, E. P. 1996. *Adv. Mater.* 8: 29.
22 Garces, J. M., et al. 2000. 12: 1835.
23 Dzenis, Y. 2008. *Science* 319: 419.
24 Nanotechwire.com, http://www.nanotechwire.com/news.asp?nid=6074 (accessed January 2010).
25 Breuer, O. et al. 2004. Polym. Compos 25: 630.
26 Nanotechnology Chemical and Engineering News, http://pubs.acs.org/n anotechnology/7842/7842business.html (accessed January 2010).
27 Boeing, http://www.boeing.com/commercial/787family/programfacts.html (accessed January 2010).
28 The MITRE Corporation, http://www.mitre.org/work/tech_papers/tech_papers_06/05_1248/05_1248.pdf (accessed January 2010).
29 Lincoln, D. M., et al. 2000. *IEEE Aerospace Conference Proceedings* 4: 183.
30 Njuguna, J., et al. 2003. *Adv. Eng. Mater.* 5: 769.
31 Sen, S., et al. 2009. *JOM* 61: 23.
32 Lake, M. L. 2001. Proceedings of the NATO Advanced Study Institute. Dordrecht, The Netherlands: Kluwer Academic 331.
33 The Hong Kong University of Science and Technology, http://www.ust.hk/eng/news/archive/e_pa060621-1583.pdf (accessed January 2010).
34 Hu, L., et al. 2007. *J. Appl. Phys.* 101: 016102.
35 Moniruzzaman, M., et al. 2006. *Macromolecules* 39: 5194.
36 Peng, H. J. 2008. *Am. Chem. Soc.* 130: 42.
37 O'Connor, I., et al. 2009. *Carbon* 47: 1983.
38 Lee, Y. K. et al. 2010. *Synthetic Met.* 160: 814.
39 Calado, H. D. R., et al. 2008. *Mater. Res. Soc. Symp. Proc.* 1143: 7.
40 Zhou, C., et al. 2008. *Carbon* 46: 1232.
41 Kumar, S., et al. 2009. *Nanotechnology* 20: 465708.
42 Shim, B. S., et al. 2007. *Chem. Mater.* 19: 5467.
43 Ramasubramaniam, R., et al. 2003. *Appl. Phys. Lett.* 83: 2928.
44 Huh, J. W., et al. 2009. *Appl. Phys. Lett.* 84: 223311.
45 Samad, M. A., et al. 2010. *Tribol. Lett.* 38: 301.
46 Li, N., et al. 2006. *Nano. Lett.* 6: 1141.
47 Yang, Y. L., et al. 2005. *Nano. Lett.* 5: 2131.

48 Al-Saleh, M. H., et al. 2009. *Carbon* 47: 1738.
49 Ray, H., et al. 2002. *Science* 297: 787.
50 Li, C., et al. 2008. *Compos. Sci. Technol.* 68: 1227.
51 Baibarac, M., et al. 2006. *J. Nanosci. Nanotechnol.* 6: 1.
52 Yoon, H., et al. 2006. *Smart Mater. Struct.* 15: 14.
53 Wei, C., et al. 2006. *J. Am. Chem. Soc.* 128: 1412.
54 Zhang, W., et al. 2006. *J. Nanosci. Nanotechnol.* 6: 960.
55 Kang, I. P., et al. 2006. *Smart Mater. Struct.* 15: 737.
56 Schulte, K., et al. 1989. *Compos. Sci. Technol.* 36: 63.
57 Weber, I., et al. 2001. *Compos. Sci. Technol.* 61: 849.
58 Kupke, M., et al. 2001. *Compos. Sci. Technol.* 61: 837.
59 Schueler, R., et al. 2001. *Compos. Sci. Technol.* 61: 921.
60 Fiedler, B., et al. 2004. *Ann. Chim. Sci. Mater.* 29: 81.
61 Thostenson, E. T., et al. 2006. *Adv. Mater.* 18: 2837.
62 Baughman, R. H., et al. 1999. *Science* 284: 1340.
63 Levitsky, I. A., et al. 2004. *J. Chem. Phys.* 121: 1058.
64 Landi, B. J., et al. 2005. *Mater. Sci. Eng. B* 116: 359.
65 Lee, D. Y., et al. 2007. *Sens. Actuators A* 133: 117.
66 Yun, Y. H., et al. 2005. *Smart Mater. Struct.* 14: 1526.
67 Liu, S., et al. 2010. *Proceedings of SPIE: The International Society for Optical Engineering* 7642: 764219.
68 Lehmann, S., et al. 2000. *Nature* 410: 447.
69 Lee, S. H., et al. 2010. *Sensors Actuators B* 145: 89.
70 Biso, M., et al. 2009. *Phys. Status Solidi. B* 246: 11.
71 Dai, C. A., et al. 2009. *Smart Mater. Struct.* 18: 085016.
72 Chen, L. Z., et al. 2008. *Appl. Phys. Lett.* 92: 263104.
73 Goo, N. S., et al. 2007. *Smart Mater. Struct.* 16: N23.
74 Park, J. H., et al. 2002. *J. Power Sources* 105: 20.
75 Park, J. H., et al. 2002. *Electrochem. Solid State Lett.* 5: H7.
76 Schadler, L. S., et al. 1998. *Appl. Phys. Lett.* 73: 3842.
77 Wagner, H. D., et al. 1998. *Appl. Phys. Lett.* 72: 188.
78 Qian, D., et al. 2000. *Appl. Phys. Lett.* 76: 2868.
79 Jurevicz, K., et al. 2001. *Chem. Phys. Lett.* 347: 36.
80 Frackowiak, E., et al. 2002. *Fuel Process Technol.* 77–78: 213.
81 Haddon, R. C. 2002. *Acc. Chem. Res.* 35: 977.
82 Journet, C., et al. 1997. *Nature* 388: 756.
83 Liu, X. M., et al. 2010. *J. Power Source*, 195: 4290.
84 Baibarac, M., et al. 2006. *Small* 2: 1075.
85 Chen, J., et al. 2007. *Chem. Mater.* 19: 3595.
86 McClory, C., et al. *Aust. J. Chem.* 62: 762.

Index

For Product Safety Concerns and Information please contact our
EU representative GPSR@taylorandfrancis.com Taylor & Francis
Verlag GmbH, Kaufingerstraße 24, 80331 München, Germany